Time-of-Flight Mass Spectrometry

Instrumentation and Applications in Biological Research

Robert J. Cotter
Middle Atlantic Mass Spectrometry Laboratory
The Johns Hopkins University
Baltimore, Maryland

ACS Professional Reference Books

American Chemical Society
Washington, DC

Library of Congress Cataloging-in-Publication Data

Cotter, Robert J., 1943–

    Time-of-flight mass spectrometry : instrumentation and applications in biological research / Robert J. Cotter.
       p. cm.— (ACS professional reference books)
    Includes bibliographical references and index.
    ISBN 0–8412–3474–4

    1. Mass spectrometry. 2. Biomolecules—Analysis. I. Title. II. Series: ACS professional reference book.
QP519.9.M3C68   1997
572′.36—dc21   97–3734
                                            CIP

The paper used in this publication meets the minimum requirements of American National Standard for Information Sciences—Permanence of Paper for Printed Library Materials, ANSI Z39.48-1984.

PRINTED IN THE UNITED STATES OF AMERICA

Dedication

This book is dedicated to my wife Catherine Fenselau, an outstanding scientist who is currently chair of the Department of Chemistry and Biochemistry at the University of Maryland Baltimore County, about eight miles from my own laboratory. Her achievements in developing biological mass spectrometry complement my own interests in instrument development, and have encouraged and facilitated the bridge between instrumentation and biology. In addition to all else, she has been my most important collaborator.

Contents

About the Author

ROBERT J. COTTER is professor of pharmacology and molecular sciences as well as professor of biophysics and biophysical chemistry at The Johns Hopkins University School of Medicine in Baltimore, Maryland. He received his bachelor of science degree in chemistry from the College of the Holy Cross in Worcester, Massachusetts, in 1965, and his doctorate in physical chemistry from The Johns Hopkins University in 1972. During his tenure at the School of Medicine, he developed a series of time-of-flight (TOF) mass spectrometers, using infrared laser desorption (IRLD), liquid secondary ion mass spectrometry (SIMS) and matrix-assisted laser desorption/ionization (MALDI), all of which utilized time-delayed ion extraction to improve mass resolution and provide more structural information for biological molecules from fragmentation in the source. More recently he developed one of the first tandem time-of-flight mass spectrometers and a curved-field reflectron for simultaneous focusing of amino acid sequence ions formed by post-source decay. Dr. Cotter's interests in instrument development also include ion trap mass spectrometry (ITMS), and he currently holds several patents on both TOF and ITMS technology. Instrument development in his laboratory has been driven by a number of ongoing collaborative projects in structural biology that include the structure and processing of the amyloid peptides implicated in Alzheimer's disease, the antigenic and anchor (lipid A) regions of bacterial lipopolysaccharides, and the amino acid sequencing of Class I viral and tumor-specific antigens aimed at the development of vaccine strategies for HIV and cancer.

Dr. Cotter is one of the founding members of Sequenom, Inc. (LaJolla, CA) involved in the development of mass spectrometry approaches to DNA sequencing, has developed time-of-flight instruments for Kratos Analytical (Manchester, England) and Shimadzu (Kyoto, Japan), and is currently developing a miniaturized "tinyTOF" mass spectrometer for bacterial agent detection in collaboration with researchers at the Johns Hopkins Applied Physics Laboratory. He is a frequent contributor to the A-pages of *Analytical Chemistry*, and edited an ACS Symposium Series volume on *Time-of-Flight Mass Spectrometry* in 1994. He is a member of the American Chemical Society and the Division of Analytical Chemistry. He is also vice president for programs (president-elect) of the American Society for Mass Spectrometry.

Dr. Cotter is married to Catherine Fenselau, professor and chair of the Department of Chemistry and Biochemistry at the University of Maryland Baltimore County, and resides in Baltimore, Maryland.

Acknowledgments

The writing of such a book is the result of many years of experience in the subject which, for this author, began with the excellent training in instrumentation that he received as a member of the research group of Walter S. Koski and Joyce Kaufmann in the Department of Chemistry at The Johns Hopkins University. In my early years at the Medical School, the successful development of time-of-flight mass spectrometry depended upon the talents and efforts of several members of my own research group, including Richard VanBreemen, Jean-Claude Tabet, Hisako Hara (Hotta), Laurence Silly, Mershid Alai, Rong Wang, Bart Emary, James K. Olthoff, William Martin, and Bernhard Spengler. In more recent years I acknowledge the contributions of Marc Chevrier, who constructed the nitrogen-based MALDI mass spectrometer, Marcela Cordero for her application of tandem TOF MS to fullerenes, and Ihor Lys, who developed the data acquisition and processing system known as *TOFware*. Three research associates continue within my group to advance this technology. Timothy J. Cornish designed both the curved-field reflectron and tandem instruments described in this book; in collaboration with Sequenom, Inc. (La Jolla, CA) he is now developing an instrument for sequencing DNA oligomers captured on large arrays of duplex probes. Vladimir Doroshenko has been responsible for adding ion trap mass spectrometry to our instrument development portfolio, including a tandem *trapTOF* instrument that will enable monoisotopic mass selection of peptides. Amina S. Woods has developed those important skills in high performance liquid chromatography, immunoprecipitation methods, and enzyme digestion chemistry that have enabled the laboratory to utilize TOF mass spectrometry to address structural problems in biology, including the sequencing of peptides associated with Alzheimer's disease, HIV, and cancer.

I also acknowledge those who have been important collaborators, including Lance Pohl and Yoichi Osawa (structure of heme proteins), Kuni Takayama and Nilofer Qureshi (lipopolysaccharides), Alex Roher (amyloid peptides associated with Alzheimer's disease), Elizabeth Jaffee, Drew Pardoll, Marc Soloski, Amy DeCloux and Robert Siliciano (Class I antigens), and Richard Benson, Wayne Bryden, and Harvey Ko with whom we are currently developing a *tinyTOF* mass spectrometer for biological agent detection. These collaborations have defined our instrument development goals, enriched our own experience in biological areas, and helped this author to bring both instrumentation and biological applications together in a single volume. Equally important have been our associations with Kratos Analytical (Manchester, UK), the Shimadzu Corporation (Kyoto, Japan), and Sequenom (La Jolla, CA). In particular I acknowledge contributions to our TOF development efforts by William Henderson, David Dingley, Brian Stall, Koichi Tanaka, Junko Iida, Kozo Shimazu, Roger Greathead, Hubert Koester, Charles Cantor, Kai Tang, Dan Little, and Nola Masterson.

A number of others deserve mention here as well. Until his untimely death in 1995, Edward Kratfel was responsible for machining and fabrication of the instruments developed in our laboratory. That task has now been assumed by his son Michael Kratfel. Much of the material and approach for this book was developed in several short courses, the latest of which were organized by Martha Vestling. The research and development, which has made our expertise possible, has been supported primarily by the National Science Foundation (NSF) and the National Institutes of

Health (NIH). Additionally, we have received considerable support and encouragement for our efforts in biological agent detection from Millie Donlon of the Defense Advanced Research Projects Agency (DARPA).

And there are also those who helped directly in the writing of this book. Considerable encouragement for the development of this volume came from Anne Wilson of ACS Books, who was involved in publishing the 1994 ACS Symposium Series volume on *Time-of-Flight Mass Spectrometry* edited by the author. She and Barbara Pralle have provided considerable assistance in the development of this book. In addition, the author is indebted to Randy Frey for the excellent design and production of the book, and to Laura Manicone for copyright/permissions support. The author is also indebted to those who read and provided suggestions for the book, including Christie G. Enke, Jackson O. Lay, and Liang Li, and to a number of helpful conversations with Scott Weinberger. Chris Enke has been a longtime friend and colleague in the development of time-of-flight mass spectrometry, and has contributed the Foreword to this volume. Finally, the author acknowledges the assistance of Darleen Stankiewicz, whose efforts in searching and cataloging references, providing figures, and keeping track of copyright requests were critical for the success of such a project.

Robert J. Cotter
December 1996
Baltimore

Foreword

Time-of-flight mass spectrometry (TOFMS) is deceptively simple in concept; a packet of ions given the same kinetic energy fly down an evacuated, field-free tube and arrive at an ion detector in the order of their mass-to-charge (m/z) values. The detector output current, measured as a function of time, can easily be converted into a mass spectrum. The physics and the instrumentation are both remarkably straightforward. The following inherent advantages of TOFMS are often listed by its advocates:

- It is much more conservative of sample than a scanning mass filter instrument; every ion in the bunch is detected.
- The wide-open flight tube with its absence of slits presents a very wide aperture to the source.
- It has no fundamental limit (other than detectability) on the range of m/z values analyzable and the m/z scale can be accurately calibrated from only two points.

Given all these advantages, and given the fact that the publication of this volume marks the 50th anniversary of the conception of TOFMS, it is fair to ask why it was submerged by other mass selection techniques for so long. The answer, of course, is that the implementation was not as simple as the concept. The process of creating the ion packet (called bunching) can result in a seriously low sample utilization duty cycle and can also limit the mass resolution available. On the detection end, the very fast electronics required for efficient collection of the detector output transients have not, until recently, been available at a reasonable cost. A variety of solutions to both these obstacles has now set the stage for the explosive development of time-of-flight instrumentation and applications that we are now experiencing. This beautifully organized and comprehensive book arrives at just the right time to be an invaluable guide and inspiration to the rapidly growing number of practitioners of this "new/old" technique.

Professor Cotter and I have had a mutual interest in developing the potential of time-of-flight mass spectrometry, though from rather different perspectives. Professor Cotter's interest has been stimulated by the inherent compatibility of TOFMS with pulsed ion sources such as laser desorption, and the high mass range that TOF readily affords. My interest has been stimulated by the uniquely high spectral generation rate and sample utilization efficiency possible with TOFMS, qualities that are most desirable for chromatographic detection. As this book demonstrates, the biomedical application of TOFMS has preceded its application in chromatographic detection. This is due, in part, to the compatibility of existing transient recorders with the repetition rate of the pulsed laser sources used, and also to the fact that ions ejected from a solid sample surface are already all headed in essentially the same direction. But more, it is due to the number of dedicated and creative researchers (including the author) who have contributed the many essential developments in instrumentation and methodology described in this volume. TOFMS is already solving problems in the areas of peptide sequencing, immunology, and combinatorial chemistry, and is poised to provided its unique capabilities to other areas of investigation including DNA sequencing. These successes in such important applications has led to a concurrent explosion in the market for TOFMS instruments.

This book will be of great value to those seeking to get abreast of this major new player in the arena of biological investigations via mass spectrometry. Its chapters describe the active areas of application, indicate current limitations, and anticipate future developments. Hopefully, these discussions will spark the imaginations of additional developers of both instrumentation and applications.

The rebirth of TOFMS is still in its early stages. The marvelous capabilities engendered by MALDI are barely in their second generation (now with remarkable mass resolution). TOF instruments based on electrospray ionization are emerging from the development lab. Commercial chromatographic detection systems based on TOFMS have recently appeared. Tandem MS machines based on TOF mass analysis for the final stage are now available and those employing TOF for both stages are receiving increasing attention in several research laboratories. A rich time of development and innovation still lies ahead. This volume will be the right-hand companion of all those who are part of this process.

Chris Enke
December 1996
Albuquerque

1

Overview and History

Mass spectrometers are analytical instruments that convert neutral molecules into gaseous ions and separate those ions according to the ratio of their mass to charge (m/z). Information from such instruments is generally displayed in a mass spectrum: a plot of relative intensity vs. m/z that can be used to deduce the chemical structure of a compound (Figure 1a). For a number of years, mass spectrometers were restricted to the analysis of volatile compounds, or those that could be made volatile by chemical derivatization, and were utilized primarily for compounds of interest to the natural-products or synthetic-organic chemist. In recent years, however, ionization techniques have been developed that produce intact molecular ions directly from samples in the liquid or solid phase, thus circumventing the requirement for volatility. A major consequence of these techniques has been the ability to analyze compounds of increasingly higher molecular mass (Figures 1.1b and 1.1c),[1,2] including peptides and proteins, oligonucleotides, and other biological macromolecules. This, coupled with the ability to extract information from very small quantities of sample, has ensured that mass spectrometers will play an increasingly important role in all areas of biological research.

Advances in ion optics, ion detection, laser technology, fast recording electronics, signal-processing techniques, and other technologies have been equally important for the development of modern mass spectrometers. This has been particularly true for the *time-of-flight* (TOF) mass spectrometer, introduced commercially more than 40 years ago. Indeed, it is only now possible to realize the high (theoretically unlimited) mass range and the high-sensitivity multichannel recording capabilities that were anticipated so many years ago, and now make this type of spectrometer an attractive instrument for biological research. While their low mass resolution continues to be of concern to mass spectroscopists, low-cost, bench-top, and user-friendly instruments are now available for the biological scientist that can provide molecular weight measurements of proteins with considerably higher accuracy than gel electrophoresis, or can be used to map tryptic digests, reveal post-translational modifications, determine the positions of disulfide bonds, assess carbohydrate hetero-geneity in glycopeptides, or provide amino acid sequences.

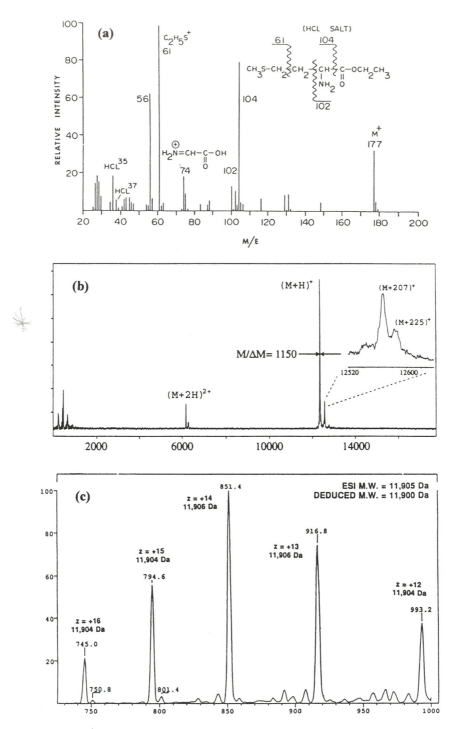

FIGURE 1.1 Examples of mass spectra. (a) Electron impact (EI) mass spectrum of methionine, which has been derivatized to its ethyl ester to improve volatility. The mass spectrum was obtained on a double-focusing sector instrument and reveals fragmentation that can be utilized to determine the structure. (b) Matrix-assisted laser desorption/ionization (MALDI), delayed-extraction mass spectrum of cytochrome c (MW = 12,360) obtained on a time-of-flight mass spectrometer. (c) Electrospray ionization (ESI) mass spectrum of gerbil muscle parvalbumin obtained on a triple-quadrupole mass spectrometer. In this spectrum, the presence of multiply charged ions enables molecular weight measurements beyond the normal mass range of the quadrupole instrument. (Figures 1.1b and 1.1c reprinted with permission, from references 1 and 2, respectively).

In the Beginning . . .

There are many different kinds of mass spectrometers, described generally by the types of ionization sources, mass analyzers, and (in some cases) detectors that are used (Table 1.1). The first mass spectrographs devised by Thomson[3] and Aston[4] used magnetic fields and recorded the resultant spatial dispersion of ions on photographic plates. In 1935, Dempster[5] described an instrument, using a combination of electric (E) and magnetic sectors (B), that was the forerunner of the double-focusing, high-resolution mass spectrometers in use today. While photoplates were used to record the mass dispersion in high-resolution instruments, it is now common to scan the magnetic field, successively bringing ions of different mass into the field of view of a particle detector. The spectra are then recorded using a computerized data system that converts the raw analog signal into convenient histogram plots, provides automatic mass calibration, and enables mass storage of multiple spectra when used in combination with gas chromatography. Although convenient, such scanning methods necessarily lose the multichannel advantage provided by photographic detection, motivating the current interest in the development of spatial array detectors. At the same time, this has provided the opportunity for recording precursor, product, and constant neutral loss (CNL) spectra using simultaneous (linked) scans of the electric (E) and magnetic (B) sectors. Other configurations include reversed-geometry (BE) for recording product ion mass spectra by scanning the electric field, triple analyzer (EBE), and four sector (EBEB) tandem mass spectrometers for high mass resolution selection of precursors and recording of the product ions formed by collision-induced dissociation (CID).

While many of the early instruments utilized thermal ionization and spark sources, the electron impact (EI) ionization source, developed by Nier,[6] provided the best method for obtaining detailed structural analysis of organic molecules. In the EI source, a solid or liquid sample is vaporized by heating, and the resultant gas molecules are ionized by a 70-eV electron beam that produces radical (odd electron) ions with high internal energy. This results in considerable fragmentation (as shown in Figure 1.1a) that can be interpreted to establish the positions of specific functional groups and the overall structure of the molecule.[7] In addition, the mass spectra of several thousand organic compounds have been recorded in databases[8] that enable rapid identification using online searches of the mass spectral library on most computerized, mass spectral data systems. This vast store of information on electron impact spectra and fragmentation patterns has been acquired primarily on sector instruments. And, because double-focusing mass spectrometers are available with mass resolutions of up to 1 part in 50,000 to 100,000, they are still considered the instruments of choice for establishing the elemental composition, as well as the structures, of organic compounds.

Mass Spectrometry Without a Magnetic Field

From very early on, mass spectroscopists have been intrigued with the idea that mass spectrometers could be constructed without magnets. In 1946, W. E. Stephens at the University of Pennsylvania suggested[9]:

> Advances in electronics seem to make practical a type of mass spectrometer in which microsecond pulses of ions are selected every millisecond from an ordinary low-voltage ion source. In travelling down the vacuum tube, ions of different M/e have different velocities and consequently separate into groups spread out in space. If the

---

TABLE I.I   Basic components of mass spectrometers.

Ionization Sources	Mass Analyzers	Detectors
Thermal ionization	Magnetic (B)	Photographic plate
Spark source	Double-focusing (EB)	Faraday cup
Electron impact (EI)	Reversed geometry (BE)	Electron multiplier
Photoionization (PI)	Ion cyclotron resonance (ICR)	Magnetic electron multiplier
Chemical ionization (CI)	Quadrupole (Q)	Continuous dynode multiplier
Field ionization (FI)	Quadrupole ion trap (ITMS)	Dual channelplate
Field desorption (FD)	Radio frequency (RF)	Daly detector
Multiphoton ionization (MPI)	Time-of-flight (TOF)	Diode array detector
Fast atom bombardment (FAB)	Fourier transform (FTMS)	Image currents
Plasma desorption mass spectrometry (PDMS)		Inductive detector
Secondary ion mass spectrometry (SIMS)	Triple quadrupole (QQQ)	
Thermospray (TS)	Four sector (EBEB)	
Infrared laser desorption (IRLD)	Hybrid (EBQQ)	
Matrix-assisted laser desorption/ ionization (MALDI)	Hybrid (EB–TOF)	
Electrospray ionization (ESI)	Tandem TOF–TOF	

Mass spectrometers can be characterized by the ionization, mass analysis, and detection methods that are used. Photographic plates were used on early mass spectrometers that employed static magnetic fields to achieve mass separation by spatial dispersion. These include the magnetic sector, double-focusing (both Mattauch-Herzog and Nier-Johnson), and reversed geometry instruments. All others are dynamic instruments that disperse ions according to frequency or time. A particularly interesting case is that of Fourier transform mass spectrometers (FTMS). In this case, excitation of ions in an ion cyclotron resonance mass spectrometer is recorded from the image currents induced on plates positioned orthogonal to the excitation signal. Not all ionization techniques have been utilized with each mass analyzer. In general, magnetic field and quadrupole mass spectrometers are most appropriately used with continuous ionization techniques, while time-of-flight mass spectrometers are most suitable for pulsed methods. The triple quadrupole, four sector, hybrid and TOF–TOF analyzers are tandem instruments. Additionally, both FTMS (ICR) and quadrupole ion trap mass spectrometers can be utilized as tandem instruments.

---

ions are collected in a fixed Faraday cage and the current amplified, then pulses of current corresponding to different M/e will be dispersed in time....This type of mass spectrometer should offer many advantages over present types. The response time should be limited only by the repetition rate (milliseconds)....Magnets and stabilization equipment would be eliminated. Resolution would not be limited by smallness of slits or alignment. Such a mass spectrometer should be well suited for composition control, rapid analysis, and portable use.

In 1948, Cameron and Eggers[10] from the Tennessee Eastman Corporation in Oak Ridge, TN, built such an instrument, which they called an *ion velocitron* (Figure 1.2). In that instrument, ions were extracted from a Nier-type electron impact source by

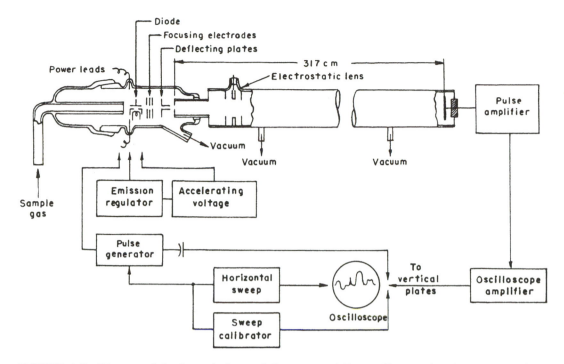

FIGURE 1.2 Diagram of the ion velocitron of Cameron and Eggers. (Reprinted with permission from reference 10).

a constant electrical field to final energies of approximately 500 eV, resulting in velocities given by:

$$v = \left(\frac{2eV}{m}\right)^{1/2}$$

[1]

that were inversely proportional to the square root of their mass/charge ratio. Their flight times were recorded at the end of a 317-cm *drift tube* and displayed on an oscilloscope (Figure 1.3). Gating of the initial ion packet was provided by a 200-V pulse applied to one of two deflecting plates placed between the ion extraction lenses and the drift tube, which also served as the trigger pulse for the oscilloscope. In their discussion, the authors noted that the deflection pulse was 5 μs wide, while the peaks observed in the mass spectrum were (in general) 20 to 30 μs in width. Thus, they suggested that the ions were not strictly monoenergetic or that they followed paths of different lengths.[10] Interestingly, while Cameron and Eggers reported the first experimental demonstration of a time-of-flight mass spectrometer, a patent for such an instrument had already been obtained by Stephens.[11] In a footnote to their paper they took note of the 1946 suggestion of Stephens, commenting that the apparatus described in this paper was near completion at that time. Thus, competition was keen even at this early stage.

In the instrument described by Cameron and Eggers, and a similar instrument reported in 1949 by Keller,[12] ions were accelerated to constant energy (eV), resulting in flight times across the drift region (D) given by:

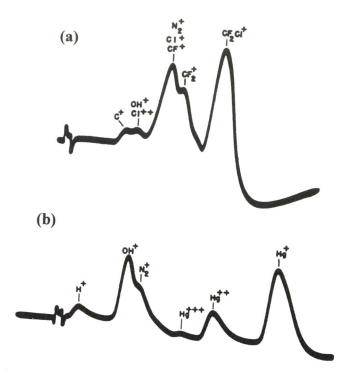

FIGURE 1.3 Electron impact mass spectra of (a) Freon-12 and (b) mercury vapor, obtained on the ion velocitron. (Reprinted with permission from reference 10).

$$t = \left(\frac{m}{2eV}\right)^{1/2} D \qquad [2]$$

that are proportional to the square root of mass. Mass resolution is given by:

$$\frac{m}{\Delta m} = \frac{t}{2\Delta t} \qquad [3]$$

and was equal to about 3 in the Cameron and Eggers instrument. In 1953, Wolff and Stephens[13] constructed an instrument in which ions were accelerated to constant momentum. In that instrument (Figure 1.4) the accelerating region consisted of a series of rings, which provided an extended linear extraction field (E). The 0.5-μs electron-beam pulse and the longer ion-extraction pulse were turned on at the same time. The extraction pulse was then turned off before the ions exited the source, and resulted in flight times given by:

$$t = \frac{Dm}{ET_p e} \qquad [4]$$

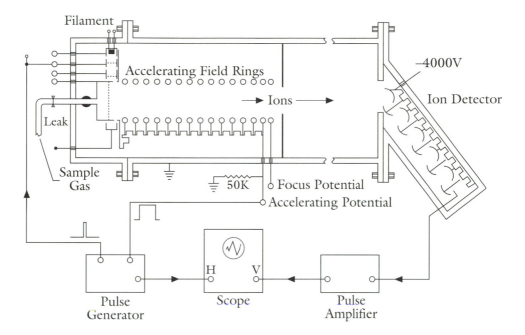

FIGURE 1.4 Diagram of the constant momentum time-of-flight mass spectrometer described by Wolff and Stephens. (Reprinted with permission from reference 13).

that produced a linear mass scale. The length of the 200-V extraction pulse (T_p) was from 5 to 50 μs, depending upon the mass range to be recorded. Mass resolution in the constant momentum case is given by:

$$\frac{m}{\Delta m} = \frac{t}{\Delta t} \tag{5}$$

and was equal to about 20 in their instrument.

In 1955, Katzenstein and Friedland[14] from the University of Connecticut described a time-of-flight mass spectrometer that was used primarily for appearance potential measurements and incorporated an electron beam that was collinear with the time-of-flight axis (Figure 1.5). Like the instrument of Wolff and Stephens, both the electron-beam and ion-extraction fields were pulsed. However, a major innovation in their instrument was that the *drawout* (or *pushout*) *pulse* used to extract the ions was activated *after* the electron beam was turned off. This enabled the ions to be formed in a field-free region. In addition, an *ion gate pulse* was added in the region between the exit of the drift tube and the detector, whose timing with respect to the drawout pulse was continuously variable. With the instrument running at a repetition rate of 5 kHz, the delay time for this 100-ns gate pulse could be scanned (manually) to produce a mass spectrum over the course of many time-of-flight cycles. This scheme was intended to address the slow response time of the detection and amplification system which, when utilized to record single transients, appeared to be the major limitation on mass resolution. A similar mass-scanning scheme had also been devised two years earlier by Ionov and Mamyrin[15] using a quadrupole gate.

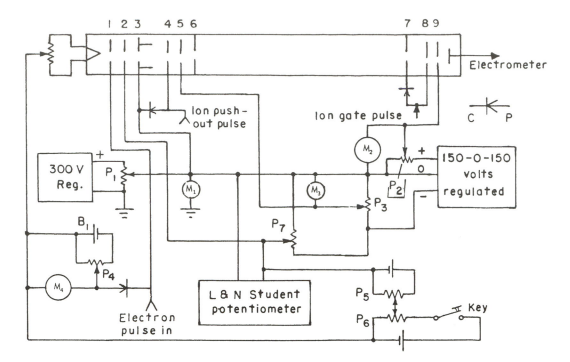

FIGURE 1.5 Diagram of the time-of-flight mass spectrometer of Katzenstein and Friedland. (Reprinted with permission from reference 14).

Time-Lag Focusing and the Bendix Mass Spectrometer

In 1955, Wiley and McLaren[16] introduced a time-of-flight mass spectrometer utilizing a two-stage, pulsed ion extraction scheme that systematically addressed the effects of the initial spatial and kinetic energy distributions of ions upon mass resolution (Figure 1.6). Ions were formed by electron impact in an *open* (and field-free) source and extracted by a low-voltage (300-V) drawout pulse that provided correction for formation of ions in different regions of the extraction field. In the second stage, ions were accelerated to their final energies of 2700 eV. The most innovative aspect of their scheme was the introduction of a time-delay between the ionization and drawout pulses. Known as *time-lag focusing*, it compensated for the initial kinetic energy distribution of the ions by allowing the ions to drift in the field-free source prior to application of the drawout pulse. The instrument (Figure 1.7) was later commercialized[17] by the Bendix Corporation (Detroit, MI), had a mass range (at a repetition frequency of 10 kHz) of about 400 amu, and a mass resolution of about 1 part in 200. While mass spectra were initially displayed and recorded using an oscilloscope (Figure 1.8), the Bendix model MA-1 used a newly developed magnetic electron multiplier (MEM) that incorporated electronic gating of the ion signal from a narrow (10–40 ns) time window that could be advanced automatically during each time-of-flight cycle. This then permitted the recording of mass spectra on a standard strip-chart recorder.

Although the Bendix time-of-flight mass spectrometers were hardly the portable instruments envisioned by Stephens, it is clear that Wiley had in mind a versatile

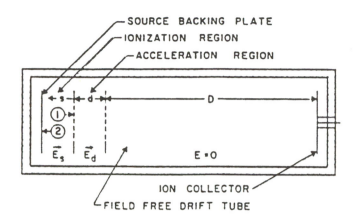

FIGURE 1.6   Basic geometry of the time-of-flight mass spectrometer of Wiley and McLaren using two-stage ion extraction and time-lag focusing. (Reprinted with permission from reference 16).

FIGURE 1.7   Photograph of the Bendix time-of-flight mass spectrometer during use by the author, background left. (Courtesy of Gettysburg College).

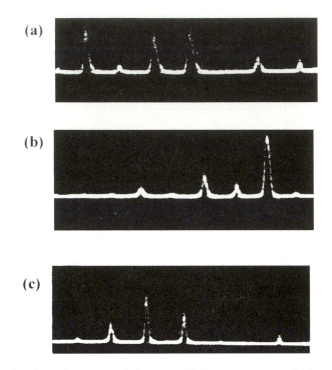

FIGURE 1.8   Oscillographic traces of the time-of-flight mass spectra of (a) xenon isotopes, (b) *n*-butane, and (c) air, obtained on the Bendix mass spectrometer. (Reprinted with permission from reference 18).

instrument, whose fast response time would enable its use as a detector for gas chromatography or as a means of following fast chemical reactions. Moreover, he shared the view of many of today's instrument developers that such instruments might be found in the laboratories of non-mass spectroscopists:

> In many cases, the samples produced in [chemistry] laboratories are analyzed by a separate mass spectrometer group, often resulting in serious delays. The simplicity and low cost of this instrument when it is equipped with only an oscilloscope output make it practical for a chemist to use it in his own laboratory as a versatile, fast-responding tool to aid him in simple analyses and the monitoring of chemical reactions.[18]

The Bendix mass spectrometer must be considered a commercial success, since many of these instruments can still be found in industrial and academic laboratories throughout the United States. In later years, manufacture of these instruments was taken over by CVC Products (Rochester, NY), which continued to produce the model 2000 until a few years ago. An attractive feature of the Bendix instruments seems to have been the ease with which they could be modified for specific applications. In 1962, Lehrle et al.[19] replaced the electron beam in the Bendix ion source with an ion beam source that could be used to generate either ions or neutrals with kinetic energies in the range of 0 to 2 keV. While used primarily for studies of gas-phase ion-molecule or neutral-molecule reactions, it was also further modified to enable sputtering of surfaces. Shortly after the introduction of chemical ionization by Field and Munson[20] in 1966, Futrell et al.[21] described a modified ion source for the Bendix instrument that could accommodate the high source pressures required by that technique. A modified instrument described by Vastola and co-workers[22] was used for the pulsed laser vaporization of coals followed by a rapid succession of electron impact time-of-flight cycles that could be recorded as a series of vertically displaced traces on an oscilloscope. However, the success of the time-of-flight mass spectrometer for organic structural analysis was severely limited by both its mass range and mass resolution. Equally important were the severe limitations on sensitivity resulting from the combination of a pulsed ionization source and the *boxcar* recording method, which together produced an extraordinarily low duty cycle. Nevertheless, it was estimated that in 1962, one-third of the mass spectrometers in use in the United States were time-of-flight instruments.[23] Accounts and reviews of this early history have been published by Harrington,[24] Pavlenko et al.,[25] and Price and Milnes.[26]

Today we appreciate that time-of-flight mass spectrometers have both unlimited mass range and the ability to record ions of all masses simultaneously, the so-called *multichannel advantage*. While oscilloscope traces could be photographed to record the analog signals, electronic recording devices that could capture, store, and integrate single transients were not available in the 1960s and 1970s. This necessitated the *scanning* approach developed on the Bendix instruments, with the consequent loss of the multichannel advantage. Furthermore, in order to recover some reasonable level of sensitivity, it was necessary to employ relatively high repetition rates (usually 10 kHz), which restricted the mass range to ions whose flight times were less than the cycle time of 100 µs. For a 1-m flight tube, this was approximately 400 daltons (Da).

At the 11th Annual Conference on Mass Spectrometry and Allied Topics in 1963, Vestal and co-workers[27] at the Johnston Laboratories in Baltimore, MD, described a *coincidence* time-of-flight mass spectrometer that recorded ions of all masses without resorting to such scanning techniques. In their instrument, positive ions and secondary electrons produced in single ionization events were extracted promptly in opposite directions and recorded in separate detectors. Detection of a secondary electron initiated a linear voltage ramp that was terminated by detection of the coincident positive ion. The voltage at the top of the ramp (determined by the ion flight time through the drift region) was converted to a pulse of proportional amplitude by a time-to-pulse-height converter. That pulse was then passed to a multichannel analyzer which, upon accumulation of many time-of-flight cycles, produced a mass spectrum. This scheme is, of course, a familiar one in many of those instruments which today incorporate time-to-digital converters, most notably the plasma desorption mass spectrometer (PDMS).

Dynamic Mass Spectrometry

The long-term success of the time-of-flight mass spectrometer was also diminished by the emergence of the quadrupole mass spectrometer which was introduced[28,29] at about the same time as the Bendix instrument. As a true *massenfilter*, mass resolution was not subject to the effects of initial kinetic energy distributions that characterized both magnetic and time-of-flight instruments, both of which required ion acceleration to a constant kinetic energy. The quadrupole mass spectrometer could achieve higher mass scanning rates than sector instruments, which was a distinct advantage when coupled with on-line with gas chromatography (GC). And, because the quadrupole mass spectrometer utilized continuous ionization, its duty cycle (and therefore its sensitivity) was considerably higher than that of the time-of-flight spectrometer. Thus, although a GC–TOF instrument was reported by Gohlke[30] in 1959, quadrupole mass spectrometers rapidly dominated the GC–MS market.

Like the time-of-flight mass spectrometer, quadrupole mass analyzers provided an alternative to instruments employing a magnetic field; and they are among a group of instruments that carry out mass dispersion according to frequency or time. Such instruments, which are known collectively as dynamic mass spectrometers, now include TOF, radiofrequency (RF), quadrupole (Q), ion cyclotron resonance (ICR), Fourier transform (FTMS) and ion trap (ITMS) mass spectrometers (Table 1.1). An interesting account of the development of these instruments is provided by Dennis Price[31] who, along with John Todd, hosted the triennial European Mass Spectrometry Symposia from 1967 to 1989 devoted to dynamic methods.[32–39]

Desorption Methods

Both electron impact (EI) and chemical ionization (CI) required that molecules be volatile, a requirement that was generally addressed by conversion of polar hydroxyl and carboxyl groups to their methoxy, methyl/acetyl ester, or trimethylsilyl derivatives, but limited the range of compounds that could be analyzed. Both are inherently continuous techniques that are used most easily on magnetic sector and quadrupole instruments. We have noted that pulsed electron-impact sources have been used with time-of-flight instruments, but that this contributed to their low duty cycle. And, although a chemical ionization source for the time-of-flight spectrometer has been described,[21] the high-pressure requirement of that technique was not particularly compatable with the space-focusing schemes that were best accomplished using an open source.

Field desorption (FD), introduced by Beckey and Schulten[40] in 1960, enabled the direct desorption/ionization of solid samples without prior derivatization. Again, because this was a continuous ionization technique that is not easily pulsed, it was used primarily on double-focusing mass spectrometers, where it could also take advantage of the higher mass ranges of these instruments. For some years, this was the most effective (albeit difficult) method for biological molecules, and was used in combination with linked scans of the magnetic and electric fields on a double-focusing mass spectrometer for the amino acid sequence analysis of peptides.[41]

In 1981, Barber and co-workers[42] at the University of Manchester Institute of Science and Technology introduced fast atom bombardment (FAB), which produced a continuous source of ions from nonvolatile molecules dissolved in a liquid matrix. Such sources were easily retrofitted to existing sector and quadrupole mass spectrometers, and they profoundly improved capabilities for the structural analysis of peptides, carbohydrates, and other biological molecules. In addition, a *continuous flow* FAB interface was later introduced[43] that enabled online interfacing with *high-performance liquid chromatography* (HPLC). Thus, this powerful, new ionization technique continued to reinforce the preeminence of sector and quadrupole mass spectrometers, and did little to advance either the technology or the popularity of time-of-flight instruments.

In the 1970s, the time-of-flight mass spectrometer reemerged in a new form. The multiple-stage, pulsed (and delayed) extraction scheme for focusing gas phase ions produced by electron impact gave way to a simpler, prompt high-voltage extraction of ions formed by short laser pulses or heavy-ion bombardment. In addition, multiplex recording devices were now available that greatly improved the duty-cycle advantage with respect to scanning instruments. In 1975, Hillenkamp and co-workers[44] introduced the laser microprobe mass analyzer (LAMMA), which was commercialized by Leybold-Hereaus (Köln, West Germany) and incorporated a Biomation (Cupertino, CA) 100 Msample/s *waveform* or *transient recorder* for capturing the analog signal produced by a single laser pulse. It also utilized a *reflectron*, an electrostatic device introduced by Mamyrin et al.[45] for improved energy focusing, and was initially intended for the analysis of atomic species from metal surfaces or biological specimens. However, following a report by Kistemaker and co-workers[46] on the ionization of intact nonvolatile compounds by infrared laser desorption (IRLD), the LAMMA, as well as a number of other laser desorption TOF instruments,[47,48] found considerable use for the structural analysis of biological compounds. At the same time, similar configurations were being developed for the simultaneous recording of ions produced by multiphoton ionization (MPI),[48] including both nonresonant and resonance enhanced (REMPI) techniques.

In 1974, Macfarlane and Torgerson[49] introduced plasma desorption mass spectrometry (PDMS), which utilized a time-to-digital converter (TDC) and had a profound effect on the role of time-of-flight mass spectrometers in biological research. This instrument was commercialized by Bio-Ion Nordic (Uppsala, Sweden) and was capable of molecular weight measurements of peptides and small proteins in the mass range up to 20 kDa.[50] In terms of both sensitivity and mass range, plasma desorption mass spectrometry achieved capabilities that could not be realized on either sector or quadrupole instruments. More than 50 Bio-Ion 20 instruments were sold worldwide; an impressive number, but considerably less than either sector or quadrupole mass spectrometers. Wide acceptance of this technique continued to be limited by its low mass resolution. In addition, the PDMS could not be easily interfaced online with HPLC or configured as a tandem instrument that could provide structural information (particularly amino acid sequences) using collision induced dissociation (CID).

Recently, two new ionization techniques were introduced that have greatly strengthened the role of mass spectrometry in biological research. Matrix-assisted laser desorption/ionization (MALDI), introduced simultaneously by Karas and Hillenkamp[51] and Tanaka et al.,[52] enabled the molecular weight determination of proteins with masses exceeding 300 kDa[53] and, because this is a pulsed technique, was most easily used on time-of-flight mass spectrometers. Ironically, the nearly simultaneous introduction of electrospray ionization (ESI) by Fenn and co-workers[54] ensured a similar (and competing) role for quadrupole mass spectrometers, which measure high molecular weights from a series of multiply charged ions whose average mass/charge ratios lie in the region of 1200 to 2000 Thomsons. When used in conjunction with the triple-quadrupole (tandem) mass spectrometer, ESI instruments can provide structural information and are easily interfaced to an HPLC. At the same time, time-of-flight instruments enjoy a distinct advantage as simple, low-cost, and compact (even desktop) instruments that are considerably more user-friendly than their differentially pumped electrospray–quadrupole counterparts.

In addition, time-of-flight mass spectrometers are being utilized in combination with endoproteinases to map tryptic and other enzymatic digests,[55,56] or to obtain amino acid sequences from the *ladder* peptides obtained by Edman[57] or carboxy and aminopeptidase digestion.[58,59] Methods have been described that enable direct transfer of proteins from SDS–PAGE (sodium dodecylsulfonate–polyacrylamide gel electrophoresis) gels to poly (vinylidene fluoride) (PVDF) membrane-coated sample supports.[60] These can be scanned by the ionizing laser beam to determine the location of protein bands (without staining), followed by in situ enzymatic digestion to provide peptide fragment masses that can be searched in a database. Time-of-flight mass spectrometers have been used for the identification of the structures and locations of post-translational modifications, providing important information on the cellular processing of proteins, and for the structural analysis of carbohydrates, glycopeptides, and glycolipids. In a number of laboratories, for example, time-of-flight mass spectrometers have been used to reveal the detailed structures of lipid-A and the core and antigenic carbohydrate regions of the lipopolysaccharides found on the cell walls of Gram-negative bacteria.[61,62] Most exciting are efforts to determine the molecular weights of oligonucleotides, where a recent demonstration of mock DNA sequencing of a short oligomer holds considerable promise for a method of sequencing the human genome.[63]

About This Book

In the past several years, there have been significant advances in the effective dynamic range, time resolution, and repetition rates of transient recorders. The recent availability of Gsample/second (and faster) digital oscilloscopes with long (4 Mbyte) recording memories has already improved our ability to obtain higher mass resolutions over the entire mass range.[64] At the same time, digitizers that can operate continuously will enable realization of the goal for mass spectrometry in the chromatographic timeframe, as described by Holland et al.[65] In addition, fragmentation (including amino acid sequence information) is now being observed using the method of *post-source decay* that has been described by Spengler et al.[66] Tandem instruments are under development, including hybrids with sector instruments,[67,68] and tandem (TOF–TOF) mass spectrometers.[69] These will enable multiplex recording of production CID mass spectra, and should ultimately provide significantly greater sensitivity than scanning, quadrupole-based tandems. New approaches to continuous ionization sources using ion storage[70,71] and orthogonal extraction[72] have been developed. Such approaches should provide an extraordinarily high duty cycle, and have already enabled the successful interfacing of electrospray ionization sources to the time-of-flight mass spectrometer.[73] The delayed extraction schemes first used in the Wiley–McLaren mass spectrometers were revisited in the 1980s for focusing ions formed by infrared laser desorption and ion bombardment, and for improving structural information from in-source decay.[74] In the 1990s, delayed extraction has been *re*-revisited for MALDI, providing extraordinary mass resolutions.[1] All of these developments are described in this book, and underscore the rapid progress that is being made in this area.

For an author attempting to describe this field, there is considerable risk in producing a volume that will be quickly outdated. For that reason, there has been a deliberate emphasis on the basic principles of ion extraction, focusing, and fragmentation, principles that should continue to be useful even as the technologies upon which this analytical technique depends continue to advance.

> Derivations of equations describing these principles have been included in the text, but are presented in highlighted sections (like this one) that may be omitted without reducing the readability of the book by those primarily interested in applications. In addition, there is a considerable amount of historical material, which should provide an interesting account of the development of this technique.

The potential for utilizing time-of-flight mass spectrometry for addressing structural problems in protein chemistry, molecular biology, immunology, and DNA sequencing also provides a focus for this volume, which emphasizes those instruments and techniques whose objectives are the effective analysis of large (generally biological) molecules. Thus, this book includes two chapters on the applications of time-of-flight mass spectrometry to the analysis of peptides, proteins, and glycoconjugates, and to DNA sequencing. Advances will continue to be made in this area as well, and so we have attempted to focus on strategies that should continue to be useful for addressing new structural problems.

References

1. Vestal, M.L.; Juhasz, P.; Martin, S.A., *Rapid Commun. Mass Spectrom.* **9** (1995) 1044–1050.
2. Hauer, C.R.; Staudenmann, W.; Kuster, T.; Neuheiser, F.; Hughes, G.J.; Seto-Ohshima, A.; Tanokura, M.; Heizmann, C.W., *Biochim. Biophys. Acta* **1160** (1992) 1–7.
3. Thomson, J.J., *Rays of Positive Electricity*, Longmans Greens (1913) London.
4. Aston, F.W., *Phil. Mag.* **38** (1919) 707.
5. Dempster, A.J., *Proc. Am. Phil. Soc.* **75** (1935) 755.
6. Nier, A.O.; Ney, E.P.; Inghram, M.G., *Rev. Sci. Instr.* **18** (1947) 398.
7. McLafferty, F.W., *Interpretation of Mass Spectra*, University Science, Mill Valley, CA (1980).
8. *EPA/NIH Mass Spectral Data Base*, Heller, S.R.; Milne, G.W.A., Eds.; NBS (1978).
9. Stephens, W.E., *Phys. Rev.* **69** (1946) 691.
10. Cameron, A.E.; Eggers, D.F., Jr., *Rev. Sci. Instr.* **19** (1948) 605.
11. Stephens, W.E., U.S. Patent 2,612,607 (1952).
12. Keller, R., *Helv. Phys. Acta* **22** (1949) 386.
13. Wolff, M.M.; Stephens, W.E., *Rev. Sci. Instr.* **24** (1953) 616.
14. Katzenstein, H.S.; Friedland, S.S., *Rev. Sci. Instr.* **26** (1955) 324.
15. Ionov, N.I.; Mamyrin, B.A., *J. Tech. Phys.* **23** (1953) 2101.
16. Wiley, W.C.; McLaren, I.H., *Rev. Sci. Instr.* **26** (1955) 1150–1157.
17. Wiley, W.C., U.S. Patent 2,685,035.
18. Wiley, W.C.; McLaren, I.H., *Science* **124** (1956) 817–820.
19. Lehrle, R.S.; Robb, J.C.; Thomas, D.W., *J. Sci. Instr.* **39** (1962) 458–463
20. Munson, M.S.B.; Field, F.H., *J. Am. Chem. Soc.* **88** (1966) 2621–2630.
21. Futrell, J.H.; Tiernan, T.O.; Abramson, F.P.; Miller, C.D., *Rev. Sci. Instr.* **39** (1968) 340–345.
22. Knox, B.E.; Vastola, F.J., *Laser Focus* **3** (1967) 15.
23. Hiroshi, T., *Koge Kagaku Dzassi* **67** (1964) 1769.
24. Harrington, D.B., in *Encyclopedia of Spectroscopy*, Clark, G.L., Ed.; Reinhold Publishing, New York, 1960.
25. Pavlenko, V.A.; Ozerov, L.N.; Rafal'son, A.E., *Soviet Physics-Technical Physics* **13** (1968) 431–436.
26. Price, D.; Milnes, G.J., *Int. J. Mass Spectrom. Ion Proc.* **99** (1990) 1–39.
27. Vestal, M.L.; Krause, M.; Wahrhaftig, A.L.; Johnston, W.H., *Proceedings of the Eleventh Annual Conference on Mass Spectrometry and Allied Topics*, May 19–24, 1963; San Francisco, pp. 358–363.
28. Paul, W.; Steinwedel, H., *Z. Naturforsch.* **8A** (1953) 448.
29. Paul, W.; Reinhard, H.P.; Von Zahn, U., *Z. Physik* **152** (1958) 143.
30. Gohlke, R.S., *Anal. Chem.* **31** (1959) 535.
31. Price, D., in *Time-of-Flight Mass Spectrometry*, Cotter, R.J., Ed.; American Chemical Society, Washington, DC (1994) 1–15.

32. *Time-of-Flight Mass Spectrometry,* Price, D.; Williams, J.E., Eds.; Pergamon Press, Oxford, England, 1969 (proceedings of the *First European Time-of-Flight Symposium*).

33. *Dynamic Mass Spectrometry,* Price, D.; Williams, J.E., Eds.; Heyden & Son, London, England, 1970, Vol 1. (proceedings of the *Second European Time-of-Flight Symposium*).

34. *Dynamic Mass Spectrometry,* Price, D.; Williams, J.E., Eds.; Heyden & Son, London, England, 1971, Vol 2. (proceedings of the *Second European Time-of-Flight Symposium*).

35. *Dynamic Mass Spectrometry,* Price, D., Ed.; Heyden & Son, London, England, 1972, Vol 3. (proceedings of the *Third European Time-of-Flight Symposium*).

36. *Dynamic Mass Spectrometry,* Price, D.; Todd, J.F.J., Eds.; Heyden & Son, London, England, 1976, Vol 4. (proceedings of the *Fourth European Symposium*).

37. *Dynamic Mass Spectrometry,* Price, D.; Todd, J.F.J., Eds.; Heyden & Son, London, England, 1978, Vol 5. (proceedings of the *Fifth European Symposium*).

38. *Dynamic Mass Spectrometry,* Price, D.; Todd, J.F.J., Eds.; Heyden & Son, London, England, 1981, Vol 6. (proceedings of the *Seventh European Symposium*).

39. Int. J. Mass Spectrom. Ion and Phys. **90** (1990), Price, D., Ed. (proceedings of the *Eighth and Ninth European Symposia*).

40. Beckey, H.D.; Schuelte, D., *Z. Instrumen.* **68** (1960) 302.

41. Matsuo, T.; Matsuda, H.; Katakuse, I.; Shimonishi, Y.; Maruyama, Y.; Higuchi, T.; Kubota, E., *Anal. Chem.* **53** (1981) 416.

42. Barber, M.; Bordoli, R.S.; Sedgwick, R.D.; Tyler, A.N., *J. Chem. Soc. Chem. Commun.* (1981) 325.

43. Caprioli, R.M.; Fan, T.; Cottrel, J.S., *Anal. Chem.* **58** (1986) 2949.

44. Hillenkamp, F.; Kaufmann, R.; Nitsche, R.; Unsold, E., *Appl. Phys.* **8** (1975) 341.

45. Mamyrin, B.A.; Karatajev, V.J.; Shmikk, D.V.; Zagulin, V.A., *Sov. Phys. JETP* **37** (1973) 45–48.

46. van der Peyl, G.J.Q.; Haverkamp, J.; Kistemaker, P.G., *Int. J. Mass Spectrom. Ion and Phys.* **42** (1982) 125.

47. VanBreemen, R.B.; Snow, M.; Cotter, R.J., *Int. J. Mass Spectrom. Ion and Phys.* **49** (1983) 35.

48. Cotter, R.J., *Anal. Chem.* **56** (1984) 485A.

49. Macfarlane, R.D.; Skowronski, R.P.; Torgerson, D.F., *Biochem. Biophys. Res. Commun.* **60** (1974) 616–621.

50. Cotter, R.J., *Anal. Chem.* **60** (1988) 781A.

51. Karas, M.; Hillenkamp, F., *Anal. Chem.* **60** (1988) 2299.

52. Tanaka, K.; Ido, Y.; Akita, S.; Yoshida, Y.; Yoshida, T., *Rapid Commun. Mass Spectrom.* **2** (1988) 151.

53. Karas, M.; Ingendoh, A.; Bahr, U.; Hillenkamp, F., *Biomed. Environ. Mass Spectrom.* **18** (1989) 841.

54. Yamashita, M.; Fenn, J.B., *J. Chem. Phys.* **88** (1984) 4451.

55. Woods, A.S.; Cotter, R.J.; Yoshioka, M.; Büllesbach, E.; Schwabe, C., *Int. J. Mass Spectrom. Ion Proc.* **111** (1991) 77–88.

56. Chen, L.; Cotter, R.J.; Stults, J.T., *Anal. Biochem.* **183** (1989) 190–194.

57. Chait, B.T.; Wang, R.; Beavis, R.C.; Kent, S.B.H., *Science* **262** (1993) 89–92.

58. Chait, B.T.; Chaudhary, T.; Field, F.H., in *Methods in Protein Sequence Analysis,* Humana Press, Clifton, NJ, 1987, 483–492.

59. Woods, A.S.; Huang, A.Y.C.; Cotter, R.J.; Pasternack, G.R.; Pardoll, D.M.; Jaffee, E.M., *Anal. Biochem.* **226** (1995) 15–28.

60. Fenselau, C.; Vestling, M., *American Laboratory* (1993) 72–78.

61. Qureshi, N.; Honovich, J.P.; Hara, H.; Cotter, R.J.; Takayama, K., *J. Biol. Chem.* **263** (1988) 5502–5504.

62. Cole, R.B.; Domelsmith, L.N.; David, C.M.; Laine, R.A.; DeLucca, A.J., *Rapid Commun. Mass Spectrom.* **6** (1992) 616–622.

63. Smith, L.M., *Science* **262** (1993) 530–532.

64. Cornish, T.J.; Cotter, R.J., *Org. Mass Spectrom.* **28** (1993) 1129–1134.

65. Holland, J.F.; Enke, C.G.; Allison, J.; Stults, J.T.; Pinkston, J.D.; Newcombe, B.; Watson, J.T., *Anal. Chem.* **55** (1983) 497A.

66. Spengler, B.; Kirsch, D.; Kaufmann, R.; Jaeger, E., *Rapid Commun. Mass Spectrom.* **6** (1992) 105–108.

67. Strobel, F.H.; Solouki, T.; White, M.A.; Russell, D.H., *J. Am. Soc. Mass Spectrom.* **2** (1990) 91–94.

68. Clayton, E.; Bateman, R.H., *Rapid Commun. Mass Spectrom.* **7** (1993) 719–720.

69. Cornish, T.J.; Cotter, R.J., *Anal. Chem.* **65** (1993) 1043–1047.

70. Grix, R.; Kutscher, R.; Li, G.; Gruner, U.; Wollnik, H., *Rapid Commun. Mass Spectrom.* **2** (1988) 83–85.

71. Chien, Benjamin M.; Michael, S.M.; Lubman, D.M., *Rapid Commun. Mass Spectrom.* **7** (1993) 837–843.

72. Dawson, J.H.J.; Guilhaus, M., *Rapid Commun. Mass Spectrom.* **3** (1989) 155–159.

73. Dodonov, A.F.; Chernushevich, I.V.; Laiko, V.V., in *Time-of-Flight Mass Spectrometry,* Cotter, R.J., Ed.; American Chemical Society, Washington, DC (1994) 108–123.

74. Cotter, R.J.; *Anal. Chem.* **64** (1992) 1027A–1039A.

2

Time-of-Flight Mass Spectrometers

The simplest time-of-flight mass analyzer (Figure 2.1a) consists of a short source–extraction region (s, usually of the order of a few centimeters), a drift region (D, from 0.5 to 4.0 m in length), and a detector. In the source region, the electrical field ($E = V/s$) is usually defined by a voltage (of the same polarity as the ions to be recorded) placed on the source backing plate, and is used to accelerate ions to constant energy. The drift region is field free and is bounded by an extraction grid and a second grid placed just before the detector, both of which are generally at ground potential. Ions cross this region with velocities that are inversely proportional to the square root of their masses. Thus, lighter ions have higher velocities and arrive at the detector sooner than heavier ions. Ion flight times generally fall in the range of 10 to 200 µs, and can be recorded by a digital oscilloscope to produce a mass spectrum (Figure 2.1b).

A number of instruments use a grounded backing plate and a voltage (of opposite polarity as the ions to be recorded) on the extraction grid. In this case, the same voltage is placed on the detector grid, and the drift region is allowed to float. Many instruments employ multiple extraction regions as well as a number of focusing elements. In some cases, one or more extraction grids (or lenses) may be pulsed, particularly if the ions are produced continuously. Additionally, time-of-flight mass spectrometers may incorporate electrostatic energy analyzers, reflectrons, or other energy-focusing devices to improve mass resolution. These are all considered in this book. However, it is generally useful to begin with instruments in which there are no additional fields that retard, deflect, turn around, or otherwise alter the ion trajectories once they enter the drift region, and in which there is a single extraction region.

The Time-of-Flight Equation and Mass Resolution

Ions may be formed in the gas phase (for example, by electron impact or multiphoton ionization), generally in the center of the source, or directly on the backing plate (by *laser* or *plasma desorption*). When formed at the surface of the

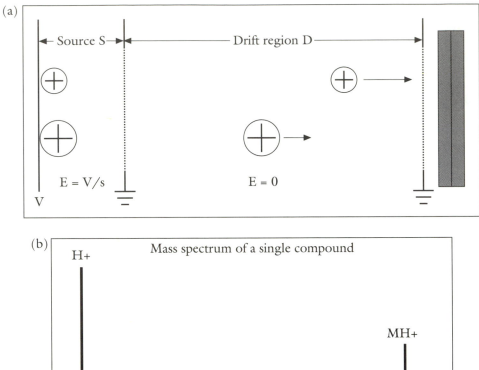

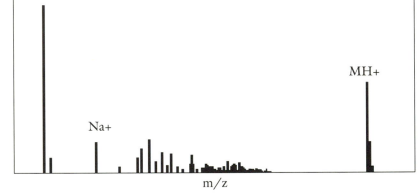

FIGURE 2.1   A time-of-flight mass spectrometer and mass spectrum.

backing plate, ions are accelerated through the entire source–extraction region to the same final kinetic energy:

$$\frac{mv^2}{2} = eV \qquad\qquad [1]$$

and cross the drift region with velocities:

$$v = \left(\frac{2eV}{m}\right)^{1/2} \qquad\qquad [2]$$

and flight times:

$$t = \left(\frac{m}{2eV}\right)^{1/2} D \qquad\qquad [3]$$

which depend upon the square root of their masses.

A more general treatment considers that ions may be formed at some distance *s* between the extraction grid and backing plate, and that ions spend a short time in the source region, where their velocities are not constant. This time (t_s) also contributes to their overall flight time, and can be calculated by considering their velocities at each position (*s*) in the source:

$$v = \left(\frac{2eEs}{m}\right)^{1/2}$$

where $v = ds/dt$, and integrating:

$$\int_0^{t_s} dt = \int_0^s \left(\frac{m}{2eE}\right)^{1/2} \frac{ds}{s^{1/2}}$$

$$t_s = \left(\frac{m}{2eE}\right)^{1/2} 2s^{1/2} = \left(\frac{m}{2eEs}\right)^{1/2} 2s$$

In addition, the flight time in the drift region (t_D) becomes:

$$t_D = \left(\frac{m}{2eEs}\right)^{1/2} D$$

and the total flight time is then given by:

$$t = t_s + t_D = \left(\frac{m}{2eEs}\right)^{1/2} [2s + D]$$

which shows the same square-root dependence upon mass. If the source region is short with respect to the drift region, and ions are formed on the backing plate, this result is essentially the same as described in equation [3].

Calibrating the Mass Spectrum. The mass scale follows a square-root law regardless of the relative sizes of the extraction and drift regions, or whether any other accelerating or decelerating regions (for example: multiple-stage extraction, Einsel lenses, or reflectrons) are utilized. Thus, mass spectra can be calibrated by measuring the flight times of two known masses (for example: H^+ and Na^+ in Figure 2.1b) to determine the constants *a* and *b* in the equation:

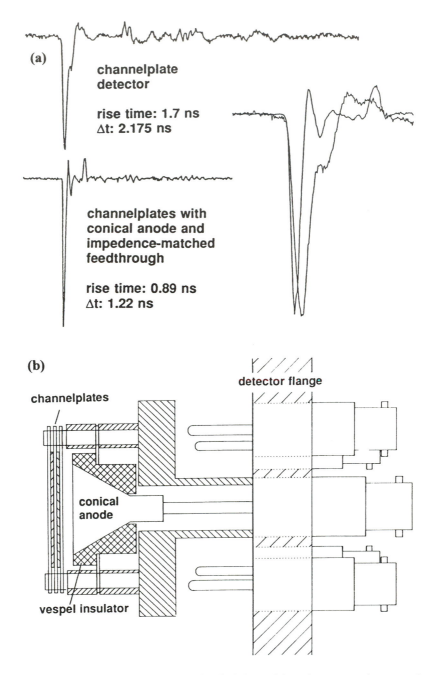

FIGURE 2.2   Signal output and design of a dual-channelplate detector with a conical anode.

$$t = am^{1/2} + b \qquad\qquad [4]$$

where b takes into account any time offsets due (for example) to laser firing time, triggering of recording devices, etc.

Mass Resolution.   In a mass spectrometer, mass resolution is defined as $m/\Delta m$. In a time-of-flight mass spectrometer in which ions are accelerated to constant energy:

$$\frac{m}{\Delta m} = \frac{t}{2\Delta t}$$ [5]

where Δt is commonly measured as the full width at half maximum (FWHM). Thus, mass resolution depends upon time resolution and (therefore) upon laser pulse widths, detector response, recorder bandwidths, and digitizing rates, as well as upon initial kinetic energies (and velocities).

The basic resolution equation is derived from rearrangement of equation [3]:

$$m = \left(\frac{2eV}{D^2}\right) t^2$$

for ions accelerated to constant energy, and its derivative:

$$dm = \left(\frac{2eV}{D^2}\right) 2t\, dt$$

leading to the result:

$$\frac{m}{dm} = \frac{t}{2dt}$$

Time, Space, and Kinetic Energy Distributions

From the very beginning, time-of-flight mass spectrometers have had a reputation as low-resolution instruments. The plasma desorption and matrix-assisted laser desorption–ionization instruments that are currently popular today produce mass spectra with resolutions from 300 to 800 that often appear to be dependent upon the nature or quantity of the sample or (in the case of MALDI) the laser power. Extraordinary mass resolutions of up to 1 part in 25,000 have been demonstrated using reflectrons,[1] but have been by no means routine. The problem is that the time-of-flight axis reflects many properties of an ion in addition to its mass, including uncertainties in the time of ion formation, its initial location in the extraction field, its initial kinetic energy (before acceleration), and metastable fragmentation.

Temporal Distributions.   These include actual distributions in the time of ion formation as well as limitations of ion-detection and time-recording devices. The classical example of the former was the 1955 instrument of Wiley and McLaren[2] in which ions were formed in the gas phase by a pulsed electron beam with a pulse width of 1 to 5 μs. In this case, temporal focusing was achieved by forming the ions in a field-free source, and then extracting these ions with a drawout pulse with a rise time of 40 ns. In the extreme case, ions are formed continuously. As discussed in Chapter 7, pulsed, orthogonal extraction enables time-of-flight mass spectra to be obtained from continuous ionization techniques such as electrospray ionization (ESI).

Distributions in the time of ion formation result in ions which enter the drift length at different times, but maintain a constant time difference (Δt) as they approach the detector. Because mass resolution is given by $t/2\Delta t$, the effects due to uncertainties in the time of ion formation can be minimized using longer flight tubes. This will have the effect of increasing t while maintaining constant Δt. The same can be said for real or apparent time distributions that arise from laser pulse widths, detector response, and recording bandwidth and digitizing rates. Figure 2.2a is a comparison of the detector response of a dual-channelplate detector with a flat anode collector and a conical anode (Figure 2.2b), obtained by recording single ions with a 400-Msample/s digitizer in the interleaving mode (effectively 4 Gsample/s or 250-ps time resolution). With the conical anode, the rise time is 0.9 ns and the peak width is 1.2 ns, which represents a Δt that is considerably smaller than the (approximately) 30-ns peak widths that are observed in most MALDI mass spectra and can be attributed to laser pulse widths (250 ps to 3 ns), initial spatial and kinetic energy distributions, space charge effects, and metastable fragmentation, all of which are discussed below. Provided that these other contributions to the peak width can be minimized, the use of 1 Gsample/s (and better) digitizers should permit higher mass resolutions to be achieved in relatively compact instruments.

Initial Kinetic Energy Distributions.   Ions are generally formed with some initial kinetic energy so that their actual flight times in the drift length are given by:

$$t = \left(\frac{m}{2KE} \right)^{1/2} D \qquad [6]$$

where $KE = eV + U_0$, and U_0 corresponds to the initial kinetic energy arising from the ion's initial components of velocity along the time-of-flight axis. As shown in Figure 2.3, ions with energies of $eV + U_0$ arrive at the detector sooner than those with no initial kinetic energy, resulting in tailing of the mass spectral peak toward the low-mass side. Although Figure 2.3 shows the velocity component of U_0 along the time-of-flight axis to be directed toward the exit of the source, it is also possible (particularly when ions are formed in the gas phase) that initial velocities for some ions will be directed away from the source exit. In the initial stages of acceleration, these ions turn around and exit the source with the same energy ($eV + U_0$) as those initially moving in the forward direction. These ions also have higher velocities and shorter flight times in the drift region, but exit the source at a later time, known as the *turn-around time*. Arriving at the detector somewhat later, they may contribute to a small amount of tailing on the high-mass side of the mass-spectral peak. Because the turn-around time contributes to a constant peak width, its effects are reduced by longer flight distances and longer flight times. The effects of initial kinetic energy spread are in general reduced by higher accelerating voltages, that is by making eV much greater than U_0.

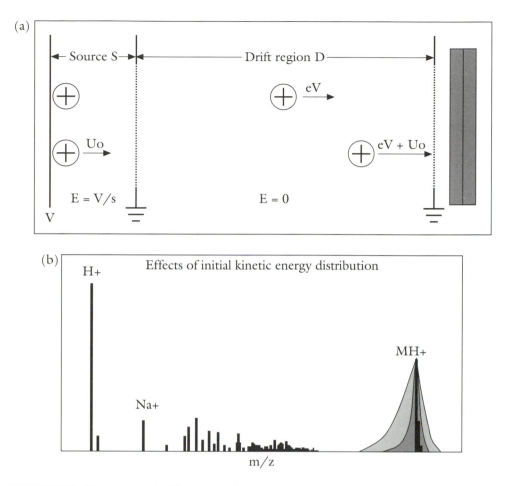

FIGURE 2.3 Two ions with different initial kinetic energies, and the effects of kinetic energy distributions on the mass spectrum. (In this and subsequent figures, the differences in the lengths of the arrows are intended to represent differences in ion velocities).

Spatial Distributions.   Ions formed in different regions of the source wil be accelerated through different distances in the extraction field, resulting (effectively) in a distribution of final kinetic energies. While this is primarily a problem for ions formed by gas-phase ionization methods, post-ionization of desorbed neutrals intersecting an incoming laser beam or thick sample layers or insulating surfaces (such as nitrocellulose) may contribute to spatial distribution effects in desorption techniques as well. Figure 2.4 compares the flight time of an ion formed directly on the surface with that of an ion formed at some distance above the surface. The ion formed above the surface exits the source region sooner, but, because its velocity is lower it arrives at the detector later. Thus, ions formed above the surface all have an energy deficit, which results in tailing of the mass spectral peak toward higher apparent mass and loss of mass resolution.

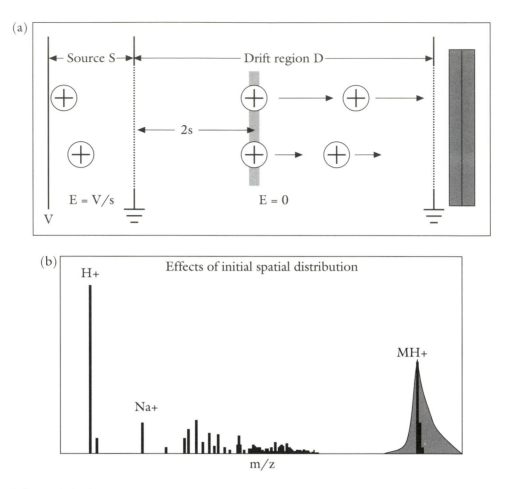

FIGURE 2.4 Two ions formed in different locations with respect to the backing plate, and the effects of spatial distributions on the mass spectrum.

When ions are formed by gas-phase methods, tailing of the mass-spectral peak is likely to be more symmetric. There is, however, a focal point, known as the space-focus plane, in the drift region at which faster ions formed toward the rear of the source catch up with slower ions formed near the front of the source. This point is independent of mass (although ions of different mass arrive there at different times) and is located at a distance 2s, where s is the full depth of the source region for desorption methods, or the distance from the extraction grid to the center of the ionization zone for gas-phase ionization methods.

As we shall see below and in the next chapter, there are two ways to correct for the loss of resolution due to the spatial distribution of ion formation. First, it is possible to devise a two-stage extraction system which pushes the space-focus plane to the entrance of the detector. The second approach considers the space-focus plane to be a virtual source with ions differing only in kinetic energy. These are then focused using a reflectron.

These effects can be demonstrated mathematically by including the initial kinetic energy (U_0) and position of ion formation (s) to calculate the final kinetic energy of an ion as it enters the drift region:

$$\frac{mv^2}{2} = U_0 + eEs$$

The final velocity again depends upon mass and is given by:

$$v = (U_0 + eEs)^{1/2}\left(\frac{2}{m}\right)^{1/2}$$

To determine the flight time in the source–extraction region we again consider that the velocity is not constant, and integrate between the time of ion formation (t_0) and the time that the ion leaves the source (t_s):

$$\int_{t_0}^{t_s} dt = \int_0^s \left(\frac{m}{2}\right)^{1/2} \frac{ds}{(U_0 + eEs)^{1/2}}$$

$$t_s = \frac{(2m)^{1/2}}{eE}\left[(U_0 + eEs)^{1/2} \mp U_0^{1/2}\right] + t_0$$

The first term $(U_0 + eEs)^{1/2}$ reflects the fact that the final kinetic energy (and velocity) depends upon the initial kinetic energy and position, but is independent of the ion's initial direction of motion. The term $\mp U_0^{1/2}$ reflects the contribution of the turn-around time, while t_0 represents uncertainties in the time of ion formation.

In the drift region the velocity is constant, so that the time spent in the drift region is given by:

$$t_D = \frac{(2m)^{1/2}D}{2(U_0 + eEs)^{1/2}}$$

and the total flight time is:

$$t = \frac{(2m)^{1/2}}{eE}\left[(U_0 + eEs)^{1/2} \mp U_0^{1/2}\right] + \frac{(2m)^{1/2}D}{2(U_0 + eEs)^{1/2}} + t_0$$

Note that the initial kinetic energy and initial position effect the flight times in both the extraction and drift regions, while the turn-around time effects only the extraction region.

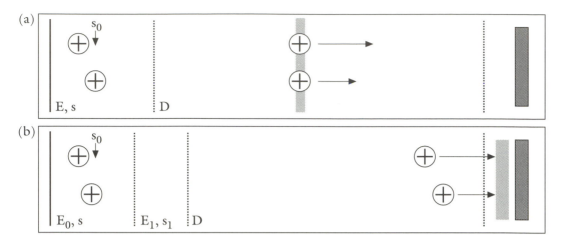

FIGURE 2.5   Comparison of single-stage and dual-stage extraction.

The Space-Focus Plane.   The early time-of-flight mass spectrometers utilized electron impact (EI) to ionize volatile molecules in the gas phase. Ionization in the gas
phase results in considerable uncertainties in the initial position of the ions in the
extraction field that are less problematic in the desorption (PDMS and MALDI)
instruments in use today. Still, it is useful to consider this problem (and the means
by which it is corrected), since spatial distributions do exist when ions are desorbed
from irregular or insulating surfaces. In addition, it is likely that compact EI–TOF
instruments will continue to be developed (and commercialized) for environmental
monitoring or as gas-chromatographic detectors.

When electron-impact ionization is used, it is common to pass the electron beam
through the center of the source region. Thus, Figure 2.5a depicts a single-stage
extraction scheme with a source length s, an extraction field $E = V/s$, and a drift
region of length D. Ions are formed at s_0 located approximately at the midpoint of
the extraction field. An ion formed at a position $s_0 + \Delta s_0$ toward the back of the
source will (as we have noted above) spend more time in the source–extraction
region, but will have higher energy and will catch up with an ion formed at s_0 at a
point d in the drift region known as the space-focus plane. The space-focus plane is
located in the drift region at the point at which ions have spent an equal time in the
extraction and drift regions:

$$\left(\frac{m}{2eEs_0}\right)^{1/2} 2s_0 = \left(\frac{m}{2eEs_0}\right)^{1/2} d \qquad [7]$$

that is, when $d = 2s_0$.

One can demonstrate that ions formed at s_0 and $s_0 + \Delta s_0$ reach d at the same
time by showing that the difference in time spent in the source region is
compensated for by the difference in time spent to reach the space-focus plane,
i.e., that $\Delta t_s = -\Delta t_d$. The difference in time spent in the source region is:

$$\Delta t_s = \left(\frac{m}{2eE}\right)^{1/2} 2\left[(s_0 + \Delta s_0)^{1/2} - s_0^{1/2}\right]$$

To solve this equation it is necessary to expand the term $(s_0 + \Delta s_0)^{1/2}$ about the point $\Delta s_0 = 0$, i.e., where the spatial distribution vanishes:

$$(s_0 + \Delta s_0)^{1/2} = s_0^{1/2} + \frac{\Delta s_0}{2s_0^{1/2}} - \frac{\Delta s_0^2}{8s_0^{3/2}} + \frac{3\Delta s_0^3}{48s_0^{5/2}} + \ldots$$

where the four terms represent the zero-, first-, second-, and third-order solutions, respectively. With a single extraction stage, only first-order solutions are possible, so that this becomes:

$$\Delta t_s \approx \left(\frac{m}{2eE}\right)^{1/2} 2\left[s_0^{1/2} + \frac{\Delta s_0}{2s_0^{1/2}} - s_0^{1/2}\right]$$

$$\Delta t_s \approx \left(\frac{m}{2eE}\right)^{1/2} \frac{\Delta s_0}{s^{1/2}}$$

Similarly, the difference in time spent to reach the space-focus plane is:

$$\Delta t_d = \left(\frac{m}{2eE}\right)^{1/2}\left[\frac{1}{(s_0 + \Delta s_0)^{1/2}} - \frac{1}{s_0^{1/2}}\right]d$$

$$\Delta t_d \approx \left(\frac{m}{2eE}\right)^{1/2}\left[\frac{s_0^{1/2} - (s_0 + \Delta s_0)^{1/2}}{s_0}\right]d$$

and if we again expand the term $(s_0 + \Delta s_0)^{1/2}$ to first order we get the result:

$$\Delta t_d \approx \left(\frac{m}{2eE}\right)^{1/2}\left[\frac{s_0^{1/2} - (s_0^{1/2} + \Delta s_0/2s_0^{1/2})}{s_0}\right]d \approx -\left(\frac{m}{2eE}\right)^{1/2}\frac{\Delta s_0}{2s_0^{3/2}}d$$

When $d = 2s_0$, this is equal to:

$$\Delta t_d \approx -\left(\frac{m}{2eE}\right)^{1/2}\frac{\Delta s_0^{1/2}}{s_0^{1/2}}$$

and $\Delta t_s = -\Delta t_d$ to first order, when Δs_0 is small.

Dual-stage Extraction and Space Focusing.   As we noted above, the location of the space-focus plane is independent of mass, while ions of different mass are focused at the space-focus plane at different times. However, mass dispersion at this short distance from the source is usually not sufficient to allow the location of a detector at this point. Thus, one solution is to move the location of the space-focus plane to a point much further from the source by using dual-stage extraction. In Figure 2.5b a dual-stage extraction scheme is depicted. This scheme moves the location of the space-focus plane to a more distant point which can be used as the location for the detector. In instruments utilizing two-stage extraction, the second extraction field (E_1) is generally much larger than the first (E_0), with larger ratios of E_1/E_0 moving the space-focus plane further from the source. Using dual-stage extraction, first-order space focusing can be achieved at almost any distance from the source, while second-order focusing is also possible for unique combinations of s_0, s_1, E_0, and E_1.

In dual-stage extraction instruments, the flight time in the source region is:

$$t_0 = \left(\frac{m}{2eE_0s_0}\right)^{1/2} 2s_0$$

where the field strength $E_0 = (V_0 - V_1)/s$. The time spent in the second extraction region is determined from its average velocity. When an ion enters this region its initial velocity is:

$$v_i = \left(\frac{2eE_0s_0}{m}\right)^{1/2}$$

Its final velocity is determined from its final kinetic energy $eE_0s_0 + eE_1s_1$:

$$v_f = \left(\frac{2e}{m}\right)^{1/2}(E_0s_0 + E_1s_1)^{1/2}$$

where $E_1 = V_1/s_1$. Thus, the time in the second region is:

$$t_1 = \left(\frac{m}{2e}\right)^{1/2}\left[\frac{2s_1}{(E_0s_0 + E_1s_1)^{1/2} + (E_0s_0)^{1/2}}\right]$$

and the time spent to reach the space-focus plane is:

$$t_d = \left(\frac{m}{2e}\right)^{1/2}\left[\frac{d}{(E_0s_0 + E_1s_1)^{1/2}}\right]$$

Again, if one considers the difference in the flight time of an ion formed at s_0:

$$t = t_0 + t_1 + t_d$$

and an ion formed at $s_0 + \Delta s_0$, and expands terms $(s_0 + \Delta s_0)^{1/2}$ to first order in Δs_0, then one can obtain the location of first-order space-focus plane shown in Figure 2.5 for any values of E_0, E_1, s_0, and s_1.

If the second extraction field (E_1) is considerably larger than the first (E_0), then an ion will spend a very short time in that region compared with the time spent in the source and drift regions. With this assumption one can obtain a rough estimate of the location of the space-focus plane that can be useful in determining the overall dimensions of the instrument.

In analogy with the single-stage space-focus equations derived above, if the time spent in the source region by an ion formed at position s_0 is:

$$t_0 = \left(\frac{m}{2eE_0}\right)^{1/2} 2s_0^{1/2} \cdot$$

then the first-order approximation for the difference in time spent by an ion formed at position $s_0 + \Delta s_0$ will be:

$$\Delta t_0 = \left(\frac{m}{2eE_0}\right)^{1/2} \frac{\Delta s_0}{s_0^{1/2}}$$

Similarly, the time it takes for by an ion formed at s_0 to travel the drift region up to the space-focus plane is:

$$t_d = \left(\frac{m}{2eE_0}\right)^{1/2} \left[\frac{1}{\left(s_0 + \dfrac{E_1}{E_0}s_1\right)^{1/2}}\right] d$$

If we let $\sigma = s_0 + (E_1/E_0)s_1$, then the difference in time between an ion formed at s_0 and one formed at $s_0 + \Delta s_0$ will be:

$$\Delta t_d = \left(\frac{m}{2eE_0}\right)^{1/2} \left[\frac{1}{(\sigma + \Delta s_0)^{1/2}} - \frac{1}{\sigma^{1/2}}\right] d$$

The series expansion of $(\sigma + \Delta s_0)^{1/2}$ is now identical to that used in the single-stage extraction described above, giving the first-order approximation:

$$(\sigma + \Delta s_0)^{1/2} \approx \sigma^{1/2} + \frac{\Delta s_0}{2\sigma^{1/2}}$$

leading to the result:

$$\Delta t_d = -\left(\frac{m}{2eE_0}\right)^{1/2} \frac{\Delta s_0}{2\sigma^{3/2}} d$$

Thus, $\Delta t_d = \Delta t_0$ when $d = 2\sigma^{3/2}/s_0^{1/2}$, and the space-focus plane can be found from the relationship:

$$d = \frac{2\left(s_0 + \dfrac{E_1}{E_0}s_1\right)^{3/2}}{s_0^{1/2}}$$

In particular, when $s_0 = s_1$, the space-focus plane is found (very approximately):

$$d = 2s_0\left(1 + \frac{E_1}{E_0}\right) \qquad [8]$$

which shows the effect of the ratio E_1/E_0 in pushing the space-focus plane down the flight tube toward the detector. Thus, if ions are formed at s_0 located at the midpoint of the source region s and accelerated to $1/10$th of their final energies in the source region, then the space-focus plane will be located at a distance from the source equal to 10 times the width of the source region.

If (in addition) $E_1 = E_0$, then $d = 4s_0$. In this case, constant field extraction occurs over the region $(s_0 + s_1)$, and our approximation reduces very nicely to the single-stage extraction case in which $d = 2(s_0 + s_1)$.

Of course, the time spent in the second extraction region is not negligible. Since all ions travel through this entire region, they all gain the same additional kinetic energy. Ions formed at $s_0 + \Delta s_0$ enter this region later than ions formed at s_0 but are already moving faster. Thus, they will travel through this region faster, and Δt_1 should be negative, reflecting the fact that this (like the drift length) is a catch-up region.

Again, the time spent in the second region for an ion formed at s_0 is:

$$t_1 = \left(\frac{m}{2eE_0}\right)^{1/2}\left[\frac{2s_1}{\left(s_0 + \dfrac{E_1}{E_0}s_1\right)^{1/2} + s_0^{1/2}}\right]$$

Substituting $\sigma = s_0 + (E_1/E_0)s_1$:

$$t_1 = \left(\frac{m}{2eE_0}\right)^{1/2}\left[\frac{2s_1}{\sigma^{1/2} + s_0^{1/2}}\right]$$

and the difference in time spent by an ion formed at s_0 and one formed at $s_0 + \Delta s_0$ is then:

$$\Delta t_1 = \left(\frac{m}{2eE_0}\right)^{1/2}\left[\frac{1}{(\sigma + \Delta s_0)^{1/2} + (s_0 + \Delta s_0)^{1/2}} - \frac{1}{\sigma^{1/2} + s_0^{1/2}}\right]2s_1$$

Again, expanding $(\sigma + \Delta s_0)^{1/2}$ and $(s_0 + \Delta s_0)^{1/2}$ to first order:

$$\Delta t_1 = \left(\frac{m}{2eE_0}\right)^{1/2}\left[\frac{1}{\sigma^{1/2} + \dfrac{\Delta s_0}{2\sigma^{1/2}} + s_0^{1/2} + \dfrac{\Delta s_0}{2s_0^{1/2}}} - \frac{1}{\sigma^{1/2} + s_0^{1/2}}\right]2s_1$$

$$\Delta t_1 = \left(\frac{m}{2eE_0}\right)^{1/2}\left[\frac{\sigma^{1/2} + s_0^{1/2} - \left(\sigma^{1/2} + \dfrac{\Delta s_0}{2\sigma^{1/2}} + s_0^{1/2} + \dfrac{\Delta s_0}{2s_0^{1/2}}\right)}{\left(\sigma^{1/2} + \dfrac{\Delta s_0}{2\sigma^{1/2}} + s_0^{1/2} + \dfrac{\Delta s_0}{2s_0^{1/2}}\right)(\sigma^{1/2} + s_0^{1/2})}\right]$$

Since the factors in the denominator are nearly equal, this becomes:

$$\Delta t_1 \approx -\left(\frac{m}{2eE_0}\right)^{1/2}\left[\frac{\Delta s_0\left(\dfrac{1}{2\sigma^{1/2}} + \dfrac{1}{2s_0^{1/2}}\right)}{(\sigma^{1/2} + s_0^{1/2})^2}\right]2s_1$$

and, since $1/2\sigma^{1/2}$ is considerably smaller than $1/2s_0^{1/2}$ when the field strength is larger in the second region, the difference in time spent in this region is:

$$\Delta t_1 = \left(\frac{m}{2eE_0}\right)^{1/2}\left[\frac{\Delta s_0}{s_0^{1/2}(\sigma^{1/2} + s_0^{1/2})^2}\right]2s_1$$

The total time difference for an ion formed at s_0 and one formed at $s_0 + \Delta s_0$ is:

$$\Delta t = \Delta t_0 + \Delta t_1 + \Delta t_d$$

$$\Delta t = \left(\frac{m}{2eE_0}\right)^{1/2}\left[\frac{1}{s_0^{1/2}} - \frac{s_1}{s_0^{1/2}(\sigma^{1/2} + s_0^{1/2})^2} - \frac{d}{2\sigma^{3/2}}\right]\Delta s_0$$

The space-focus plane is the point at which $\Delta t/\Delta s_0 = 0$, that is, when:

$$d = 2\sigma^{3/2}\left[\frac{1}{s_0^{1/2}} - \frac{2s_1}{s_0^{1/2}(\sigma^{1/2} + s_0^{1/2})^2}\right]$$

with $\sigma = s_0 + (E_1/E_0)s_1$.

When designing a time-of-flight instrument with dual-stage extraction, one first chooses values for the drift length (D) and the lengths of the source (s) and extraction (s_1) regions. These then uniquely determine the ratio E_1/E_0. If one then chooses a

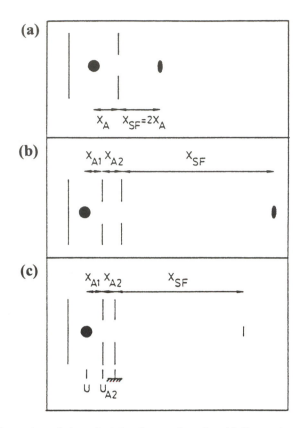

FIGURE 2.6 Illustration of the principle of space focusing. (a) first-order space focusing using single-stage ion extraction, (b) first-order space focusing using dual-stage extraction, and (c) second-order space focusing. (Reprinted with permission from reference 3).

value for the accelerating voltage (V_0 on the backing plate when a grounded flight tube is to be utilized), then this will uniquely determine the voltage (V_1) on the first extraction grid. The voltage V_1 can then be adjusted experimentally to provide the best focus.

Second-Order Space Focusing.   For arbitrary combinations of extraction fields, the space-focus plane will be a first-order one, which will be sufficient to focus ions with a minimal initial spatial distribution. However, dual-stage extraction configurations can be designed to produce a second-order space-focus plane to improve focusing and accomodate broader initial spatial distributions. Figure 2.6 (from Schlag et al.[3]) compares the focusing effects of single-stage, dual-stage, and second-order extraction. In the latter, the location of the second-order space-focus plane is relatively close to the source, where mass dispersion is low. Thus, the second-order space-focus plane is more appropriately used as the focal point for a reflectron (where the initial spatial distribution has been converted to a kinetic energy distribution), than as the location for a detector.

Schlag et al.[3,4] have obtained second-order solutions for the flight times of spatially distributed ions accelerated in a dual-stage extraction system. Using the notation shown in Figure 2.6, their relationship between the distances x_{A1}, x_{A2}, and x_{SF} is:

$$x_{A1} = \frac{x_{SF} - 2x_{A2}}{2(x_{SF} + x_{A2})}\left[x_{SF}\left(\frac{X_{SF} - 2x_{A2}}{3x_{A2}}\right) + x_{A2}\right]$$

while the relationship between the voltages U and U_{A2} is expressed by:

$$U_{A2} = \frac{U(2x_{SF} + 2x_{A2})}{3x_{SF}}$$

Dual-stage extraction is not always desirable. It utilizes relatively low extraction fields in the first stage, while focusing of ions with large initial kinetic-energy distributions is best accomplished using very high extraction fields. In addition, increasing the dimensions to achieve second-order focusing at longer distances appropriate for location of a detector also increases the turn-around time, since the field strength in the first region will be lower. Second-order focusing may be most appropriate for instruments in which ions are formed in the gas phase, since these ions are generally broadly distributed. In such instruments, the space-focus plane provides an excellent virtual source for a reflectron, since (at that point) the spatial distribution has been converted to a kinetic energy distribution. For desorption techniques (such as PDMS and MALDI), the formation of ions directly on (or near) an equipotential surface virtually eliminates the spatial distribution, so that single-stage, high-field extraction may be more appropriate.

Time-Lag Focusing

The Wiley and McLaren instrument reported in 1955 utilized pulsed, time-delayed, two-stage extraction as an approach to simultaneous time, space, and energy focusing.[2] As shown schematically in Figure 2.7, $s_0 = 0.2$ cm, $s_1 = 1.2$ cm, and the two extraction fields ($E_0 = 320$ V/cm and $E_1 = 1280$ V/cm) were determined by setting the first and second extraction grids to -64 V and -1600 V, respectively. During ionization by electron impact, the source region is field-free, since both the backing plate and first extraction grid are at ground potential. Following the ionization period, the first grid is pulsed, and ions are extracted to their final kinetic energies of 1600 eV.

Time focusing was accomplished by using an extraction pulse with a fast rise time (40 ns). In this instrument, the uncertainties in time of ion formation (Δt_0) arise from the broad ionization period of the electron-beam pulse (about 1 to 5 μs). *Space focusing* is shown for the first two ions in Figure 2.7, which have identical initial kinetic energies but are formed at different locations in the source. Using this particular set of parameters for s_0, s_1, E_0, and E_1, they determined that the space-focus

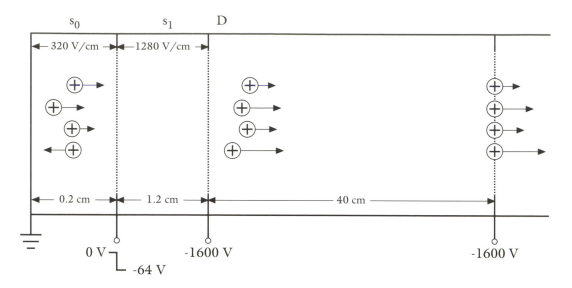

FIGURE 2.7   Illustration of the principle of time-lag focusing.

plane would be about 40 cm from the second extraction grid. Thus, the flight-tube length was set at that distance. As noted above, space focusing is essentially independent of mass.

Using the equation derived above:

$$d = 2\sigma^{3/2}\left[\frac{1}{s_0^{1/2}} - \frac{s_1}{s_0^{1/2}(\sigma^{1/2} + s_0^{1/2})^2}\right]$$

the value of σ in the original Wiley-McLaren instrument was:

$$\sigma = s_0 + \frac{E_1}{E_0}s_1 = 0.2 + \frac{(1280)}{(320)}1.2 = 5$$

The space-focus plane can be calculated:

$$d = 2(5)^{3/2}\left[\frac{1}{(0.2)^{1/2}} - \frac{1.2}{2(5)^{1/2}(5^{1/2} + 0.2^{1/2})^2}\right]$$

$$d = 22.36\ (2.236 - 0.373) = 41.6\ cm$$

resulting in space focusing at the detector.

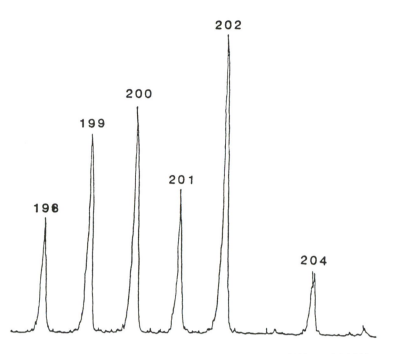

FIGURE 2.8   Mass spectrum of the isotopes of Hg obtained on a CVC model 2000 time-of-flight mass spectrometer. (Reprinted with permission from reference 5).

Initial kinetic energy focusing was achieved by a method which Wiley and McLaren introduced known as time-lag focusing. In this approach, a time delay is introduced between the end of the ionization period and the ion-extraction (or drawout) pulse. This enables the ions to drift within the field-free source before the extraction voltage is applied. In Figure 2.7 the first and third ions are formed in the same location but have different initial kinetic energies. The first ion drifts toward the extraction grid faster than the third ion. Thus, it receives less kinetic energy from the extraction field, and both ions arrive at the detector at the same time. The necessity for correcting the turn-around problem can be appreciated by comparing the first and fourth ions, which have the same initial kinetic energy but have initial velocities in opposite directions. Because the fourth ion drifts toward the back of the source prior to application of the drawout pulse, it receives considerably greater kinetic energy from the extraction field, and also catches up at the detector.

Perhaps the greatest drawback of time-lag focusing is that it is mass-dependent, that is, only a narrow range of mass can be energy focused for a particular value of the time delay. As we shall describe below, the Bendix mass spectrometers based upon Wiley and McLaren's design utilized a scanning technique that recorded only a small *time slice* during each time-of-flight cycle, and was extremely wasteful of the available data. At the same time, it provided the opportunity to link the time delay to the scanning circuit so that its length increased with recorded mass. Nevertheless, these instruments provided (for their time) extraordinary mass resolutions. Figure 2.8 shows a time-of-flight mass spectrum of the isotopes of mercury with a mass resolution of 1 part in 810. The mass spectrum was obtained on a CVC Products

(Rochester, NY) model 2000 mass spectrometer with a 2-m flight tube and multiple-stage extraction using four equally spaced grids with voltages of −150 V (pulsed), −300 V (pulsed), −1350 V, and −2700 V.[5] Note that this is effectively a dual-stage extraction scheme, since the field strengths in the first two regions and in the last two regions are identical.

Desorption from Surfaces

Much of the current interest in time-of-flight mass spectrometers is driven by instruments which desorb nonvolatile molecules (particularly peptides and other biological molecules) from surfaces. These methods include plasma desorption mass spectrometry (PDMS), laser desorption (LD), and matrix-assisted laser desorption/ionization (MALDI), and they greatly simplify the design of time-of-flight mass spectrometers, since they effectively eliminate both the time- and spatial-distribution problems.

In the plasma desorption technique, ions are formed by a 1-MeV nucleon particle impacting the surface, and are desorbed within 10^{-9} seconds of the impact. The lasers commonly used in LD and MALDI instruments generally have pulse widths in the range of 3 to 10 ns, while inexpensive pulsed nitrogen lasers with pulse widths of 250 to 600 ps are also available. Ionization times are, therefore, considerably shorter than the rise time of the drawout pulse in the Wiley and McLaren instrument, and greatly reduce the initial temporal distribution (Δt_0). In addition, because ions are desorbed from a well-defined, equipotential surface (in effect, the backing plate), the initial spatial distribution (Δs_0) is also minimized, and the flight time is given by:

$$t = \frac{(2m)^{1/2}\left[(U_0+eV)^{1/2} \mp U_0^{1/2}\right]s}{eV} + \frac{(2m)^{1/2} D}{2(U_0+eV)^{1/2}} \qquad [9]$$

where all ions receive the same energy (eV) because of acceleration. In addition, if the drift tube length D is much longer than the source length s, then the flight time of an ion with $KE = U_0 + eV$ is:

$$t = \frac{(2m)^{1/2} D}{2(U_0+eV)^{1/2}} \qquad [10]$$

In this case, it should be obvious that flight times will be relatively insensitive to initial kinetic energies (U_0) if the accelerating voltage (V) is very large.

From equation [10] the difference in arrival time for an ion with initial kinetic energy U_0 and an ion with no initial kinetic energy is:

$$\Delta t = D\left(\frac{m}{2}\right)^{1/2}\left[\frac{1}{(eV)^{1/2}} - \frac{1}{(eV+U_0)^{1/2}}\right]$$

Since mass resolution is $\Delta m/m = 2\Delta t/t$, the mass resolution when $eV \gg U_0$ becomes:

$$\frac{\Delta m}{m} \approx 2\left[\frac{(eV + U_0)^{1/2} - (eV)^{1/2}}{(eV)^{1/2}}\right]$$

Again, if we expand $(eV + U_0)^{1/2}$ to first order:

$$\frac{\Delta m}{m} \approx 2\left[\frac{(eV)^{1/2} + \dfrac{U_0}{2(eV)^{1/2}} - (eV)^{1/2}}{(eV)^{1/2}}\right]$$

and:

$$\frac{\Delta m}{m} \approx \frac{U_0}{eV}$$

This is an interesting result in that mass resolution is improved by using very high accelerating voltages, i.e., $eV \gg U_0$, and (provided that $D \gg s$) mass resolution is not dependent on or improved by longer flight-tube lengths. Most PDMS and MALDI instruments in fact utilize accelerating voltages in the range of 20 to 30 kV, while a 45-kV instrument has been reported by Chen et al.[6] Assuming that initial kinetic energies are of the order of 10 eV, one would expect mass resolutions of 1 part in 2000 to 3000. However, such resolutions are rarely achieved on linear instruments.

For ions formed by matrix-assisted laser desorption (MALDI), the initial kinetic energies may be considerably higher than 10 eV. In that technique, peptides or proteins are dissolved in a solution of a low-molecular-weight UV chromophore, such as nicotinic acid, caffeic acid, or sinnapinic acid. Independent studies by Beavis and Chait,[7] Huth-Fehre and Becker,[8] and Pan and Cotter[9] have suggested that the analyte ions are entrained in the rapid, high-velocity expansion that occurs as the matrix ions are desorbed. Assuming that ions from the matrix have average initial kinetic energies of 1 eV, the larger-mass molecular ions from peptides and proteins will have average kinetic energies that scale (roughly) with their increasing mass.

More fundamentally, the relatively low mass-resolution results observed on linear instruments suggest that desorption techniques do not entirely eliminate the problems associated with initial time and space distributions. Detector response times and digitizer sampling rates will in fact limit mass resolution in the short time intervals that are measured on high-voltage instruments, and contribute (essentially) to uncertainties in Δt_0. In addition, desorption of neutral species (as well as ions) from a 3- to 10-ns laser might well result in ionization above the surface, resulting in unanticipated spatial-distribution problems.

Fragmentation and Mass Resolution

In mass spectra, peaks corresponding to fragments of the molecular ion provide a wealth of structural information on the compound that is being analyzed. For small organic molecules, fragment peaks provide information on the identity and arrangement of functional groups. For peptides, they may be utilized to determine the amino acid sequence. Fragmentation results from the excess internal energy deposited in the molecular ion during the ionization process, and occurs at rates that depend upon the amount of excess energy, the number of degrees of freedom in the molecular ion, and the strengths of particular bonds. For small molecules with high internal energy, fragmentation may be very prompt; that is, occuring shortly after ionization. For larger molecules, such as peptides and proteins, the larger number of bonds available to partition the internal energy may extend the fragmentation time frame considerably. Because time-of-flight mass spectrometers generally begin to extract ions as soon as they are formed, fragmentation processes may extend all the way from the time of ionization to the time that ions reach the detector. The effects on the mass spectra (particularly on mass resolution) and the amount of information that is obtained depend upon where and when fragmentation occurs.

Prompt fragmentation is that which is indistinguishable in time and location from the ionization event. In desorption methods such fragmentation takes place on (or very near) the surface; for gas-phase ionization methods it occurs within the ionization zone. As shown in Figure 2.9a, fragment ions that are formed *promptly* are accelerated to the same final kinetic energies as molecular ions, so that their flight times follow the same square-root law.

Metastable fragmentation, occuring in the source while the ion is being accelerated by the field, is the most detrimental to the mass-spectral resolution. As shown in Figure 2.9b, fragmentation occuring at this time produces ions that arrive at the detector at times that are intermediate between that of the molecular ion and its fragments. These ions contribute to tailing of the molecular-ion peak and increased baseline noise.

Post-source fragmentation results from decompositions in the drift region. The resulting fragments have the same velocities as their precursors and (as also shown in Figure 2.9c) the same flight times. While this results in loss of fragment-ion information, it may be an advantage in determining the molecular weight of a protein or large peptide available in very small quantities. In addition (as will be described in Chapter 8), it is possible to recover structural information from the fragmentation that occurs in the drift region on instruments equipped with a reflectron.

In designing time-of-flight mass spectrometers, the fragmentation behavior of the compounds to be analyzed may be a more important consideration than the initial conditions. For example, dual-stage extraction might improve mass resolution for volatile organic molecules that are ionized by electron impact; however, the initial weak extraction field in a laser desorption instrument would cause more peptide ions to fragment during acceleration. In this case, one would want to minimize the time that the ions spend in the extraction field by using single-stage extraction. In an extreme case, an instrument might be designed that is intended for the sequencing of oligonucleotides using *ladder* mixtures. Because the sequence is to be obtained from a set of molecular ions, it is necessary that the amount of fragmentation occuring before the ions enter the drift length be reduced. Thus, such an instrument will need the highest possible extraction field.

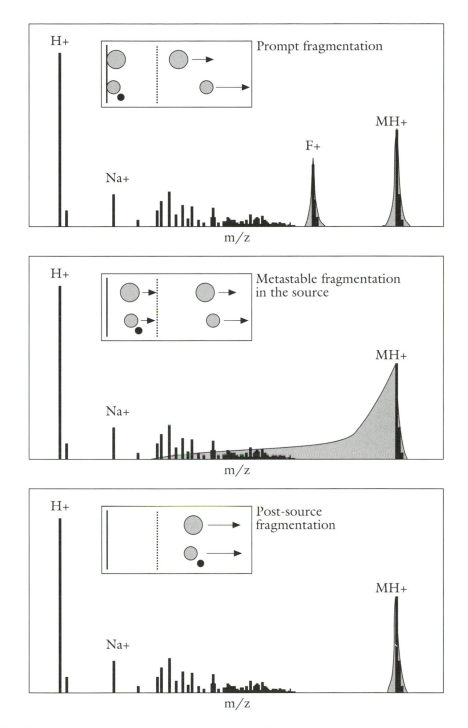

FIGURE 2.9 Comparison of the effects on time-of-flight mass spectra of prompt, metastable, and post-source fragmentation.

Recording Techniques and the Multichannel Advantage

As a last topic in this chapter, we consider the techniques that have been used to record time-of-flight mass spectra, and the unique ability of the time-of-flight mass spectrometer to record all ions of all masses without scanning.

Oscillographic Recording.   Because flight times are of the order of a few microseconds, the most common approach in the early mass spectrometers described in Chapter 1 was to photograph the traces on an oscilloscope that was triggered by the ionization or ion drawout pulse. These instruments were generally operated in a repetitive mode as a means for intensifying the oscillographic trace, but did not provide a means for capturing and storing individual time-of-flight transients which could subsequently be added and signal averaged to provide high dynamic range. Oscillographic recording did, however, provide the first demonstration of the so-called *multiplex recording advantage*: the ability to record ions of all masses without scanning.

Boxcar Recording.   The Bendix mass spectrometer that was manufactured according to the Wiley and McLaren design utilized a unique magnetic electron multiplier (MEM) that is shown in Figure 2.10. Oscillographic recording could be obtained by monitoring the output of the multiplier anode. However, the unique feature of this detection system was the ability to collect the ion signal from a small (10 to 40 ns) window from each transient, and to reconstruct the mass spectrum by recording the signals from successive intervals from a large number of transients on a conventional strip-chart recorder. This approach is commonly known as *boxcar recording* and is an effective approach for repetitive and highly reproducible signals. It was not, however, a viable approach for time-varying transients, and therefore competed poorly with quadrupole instruments at a time when there was growing interest in combined gas chromatography–mass spectrometry (GC–MS). In addition, the duty cycle (and hence the sensitivity) was very low.

The Bendix mass spectrometer had a repetition rate of 10 kHz, so that individual transients were about 100 µs in length. With an accelerating voltage of 2700 V and a drift length of 1 m, this restricted its mass range to about 400 Da (giving rise, of course, to its reputation as a low-molecular-weight instrument). If 10 ns of data are collected for each 100-µs transient, then a minimum of 10,000 transients are required to reconstruct a mass spectrum, leading to a recording duty cycle of 10^{-4}. (In practice, more cycles were utilized by incrementing the 10-ns window in each cycle by smaller steps in order to match the time response of the recorder, but this does not alter the duty cycle). In each 100-µs recording cycle, the ionization period was generally about 1 µs, resulting in an ionization duty cycle of 10^{-2}. Thus, the overall duty cycle was about 10^{-6}. Compared with quadrupole instruments, which utilized continuous ionization and (at the time) could scan the mass range from 1 to 1000 Da in 1 s with an overall duty cycle of 10^{-3}, the time-of-flight mass spectrometer proved to be a poor competitor.

Time-Slice Vs. Time-Array Detection.   Holland et al.[10] have referred to this method of data collection as *time-slice recording*. It is shown schematically in

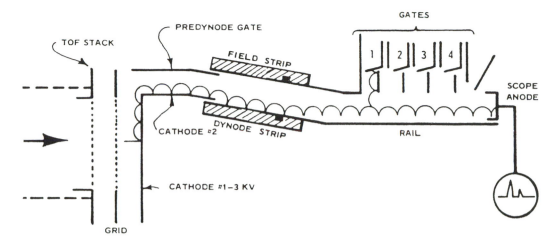

FIGURE 2.10 Schematic of the magnetic electron multiplier detection system from the Bendix time-of-flight mass spectrometer. Photographic recording was obtained by monitoring the scope anode, while the boxcar recording method used 10-ns pulses applied to one or more gates that directed a small time slice of the amplified signal to separate collectors.

Figure 2.11a, where the data collection window is a small portion of each transient.[11] In contrast, time-array detection (Figure 2.11b) refers to methods which capture and store the contents of each and every transient. For methods such as GC–MS or LC–MS, which introduce sample into the mass spectrometer continuously, the duty cycle is limited only by the ionization duty cycle which, at 10^{-2}, is far more favorable than that which can be obtained on scanning (sector or quadrupole) instruments that utilize continuous ionization. The duty cycle for time-of-flight instruments utilizing continuous sample introduction and ionization can be raised to nearly 100% using a variety of ion-storage techniques. These include the *ion well* technique described by Wollnik et al.,[12] quadrupole ion trapping by Lubman et al.,[13] and the orthogonal extraction approaches described by Guilhaus et al.[14] and Dodenov et al.[15] for electrospray ionization. For pulsed ionization techniques, such as PDMS and MALDI, sample consumption is not continuous and occurs only during ionization. For these methods, time-array detection provides the opportunity to collect all of the ions formed.

Time-array detection can be realized in two very distinct forms. Timed interval recording[10] is utilized in instruments in which a few single ions are recorded in each time-of-flight cycle, and the mass spectrum is constructed from the ion signal obtained over many time-of-flight cycles. This method is used in plasma desorption mass spectrometry (PDMS)[16] and the static SIMS instruments developed by Standing et al.,[17] where ions are generally recorded using a time-to-digital converter (TDC). This approach is discussed in more detail in Chapter 4 (Plasma Desorption Mass Spectrometry). Time amplitude measurements are used when analog signals are recorded in each time-of-flight cycle, and repetitive cycles are utilized to improve the dynamic range. Thus, laser desorption and MALDI instruments (Chapter 6) utilize transient recorders or digital oscilloscopes to record mass spectra.

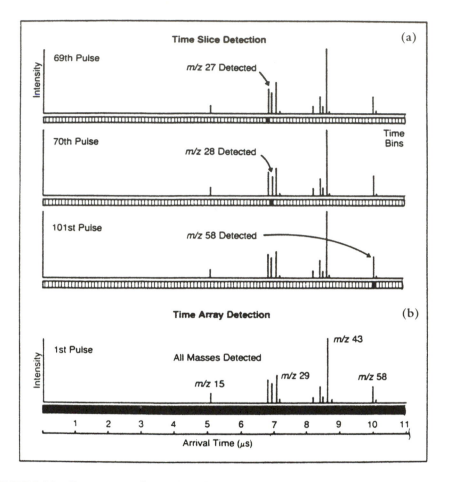

FIGURE 2.11 Comparison of time-slice detection and time-array detection. (Reprinted with permission from reference 10).

References

1. Grix, R.; Kutscher, R.; Li, G.; Gruner, U.; Wollnik, H., *Rapid Commun. Mass Spectrom.* **2** (1988) 83–85.
2. Wiley, W.C.; McLaren, I.H., *Rev. Sci. Instr.* **26** (1955) 1150–1157.
3. Boesl, U.; Weinkauf, R. Schlag, E.W., *Int. J. Mass Spectrom. Ion Processes* **112** (1992) 121–166.
4. Weinkauf, R.; Walter, K.; Weickhardt, C.; Boesl, U.; Schlag, E.W., *Z. Naturforsch., Teil A*, **44** (1989) 1219.
5. Cotter, R.J., *Biomed. Environ. Mass Spectrom.* **18** (1989) 513–532.
6. Tang, K.; Allman, S.L.; Chen, C.H., *Rapid Commun. Mass Spectrom.* **6** (1992) 365–368.
7. Beavis, R.C.; Chait, B.T., *Chem. Phys. Lett.*, **181** (1991) 479.
8. Huth-Fehre, T.; Becker, C.H., *Rapid Commun. Mass Spectrom.* **5** (1991) 378.
9. Pan, Y.; Cotter, R.J., *Org. Mass Spectrom.* **27** (1992) 3–8.
10. Holland, J.F.; Enke, C.G.; Allison, J.; Stults, J.T.; Pinkston, J.D.; Newcombe, B.N.; Watson, J.T., *Anal. Chem.* **65** (1983) 997A–1112A.

11. Holland, J.F.; Allison, J.; Watson, J.T.; Enke, C.G., in *Time-of-Flight Mass Spectrometry*, Cotter, R.J., Ed.; American Chemical Society, Washington, DC (1994) pp 157–176.

12. Grix, R.; Kutscher, R.; Li, G.; Gruner, U.; Wollnik, H., *Rapid Commun. Mass Spectrom.* **2** (1988) 83–85.

13. Chien, B.M.; Michael, S.M.; Lubman, D.M., *Rapid Commun. Mass Spectrom.* **7** (1993) 837–843.

14. Dawson, J.H.J.; Guilhaus, M., *Rapid Commun. Mass Spectrom.* **3** (1989) 155–159.

15. Dodonov, A.F.; Chernushevich, I.V.; Laiko, V.V., in *Time-of-Flight Mass Spectrometry*, Cotter, R.J., Ed.; American Chemical Society, Washington, DC (1994) 108–123.

16. Macfarlane, R.D.; Torgerson, D.F., *Science* **191** (1986) 920.

17. Chait, B.T.; Standing, K.G., *Int. J. Mass Spectrom. Ion and Phys.* **40** (1981) 185.

Reflectrons and Other Energy-Focusing Devices

In designing any time-of-flight mass spectrometer, the first opportunity to improve mass resolution is to eliminate or minimize the initial conditions (time, space, and kinetic energy distributions) which give rise to losses in resolution. As we have seen, temporal distributions can be reduced by using short laser pulses, while spatial distributions are minimized by forming ions on a surface, or in the gas phase with a tightly focused photoionizing laser beam. Initial kinetic energy distributions are perhaps the hardest to eliminate, because they are so often inherent in the ionization process itself. Kinetic energy spreads for ions formed by electron impact are actually rather small, compared (for example) with the distributions for ions formed by matrix-assisted laser desorption or electrospray ionization.

Ion extraction provides the second opportunity for improving mass resolution by compensating for the effects of initial conditions. Again as we have seen, the distribution in time of ion formation can be corrected by pulsed extraction with very fast rise time, while dual-stage extraction optics can be designed to minimize the effects of spatial distributions by pushing the space-focus plane down the flight tube toward the detector. The effects of kinetic energy distributions are minimized by using high extraction fields, so that eV will be much larger than U_0. However, no combination of static electrical fields for ion extraction can provide simultaneous focusing of the spatial and kinetic energy distributions. Thus, Wiley and McLaren[1] utilized a pulsed, time-delayed extraction scheme, which was (unfortunately) mass-dependent and, therefore, impractical for instruments simultaneously recording the entire mass range. Other time-dependent approaches have been suggested, and include impulse-field focusing,[2,3] velocity compaction,[4] dynamic field focusing,[5] and post-source pulse focusing.[6,7]

However, static fields can be used when either an initial spatial or kinetic energy spread is the dominant initial distribution. For desorption methods such as PDMS, pulsed SIMS, and laser desorption, the spatial problem is eliminated by formation of the ions on a smooth, equipotential surface, while the kinetic energy distributions (particularly in the case of MALDI) can be rather severe. In instruments in which spatial distributions dominate, dual-stage extraction (in effect) converts this to a kinetic energy distribution at the space-focus plane, that is, it actually increases the kinetic energy spread! In both cases, we are left with the task of compensating for a distribution in kinetic energy.

Thus, attention has always been focused on methods for compensating for the kinetic energy distribution of ions *after they have entered the flight tube*. This provides a third opportunity for improving mass resolution and generally involves reflection of the kinetic energy distribution back upon itself by reversing the direction of ion motion. Unlike the electrostatic energy analyzers used in double-focusing sector instruments, the reflectron does not in fact correct, or even reduce, the kinetic energy spread. However, it does compensate for the kinetic energy distribution by ensuring that ions with differing energies (but the same mass) will arrive at the detector at the same time. In effect, it reproduces the ion packet (Δt) at the source or space focus plane, but at a much longer distance and at a later time (t), so that the mass resolution ($m/\Delta m = t/2\Delta t$) improves.

The Mamyrin Reflectron

In 1973 Mamyrin and co-workers[8] published a design for a high resolution time-of-flight mass spectrometer that included a reflectron for kinetic energy focusing. In that instrument (shown in Figure 3.1) ions were formed by electron impact, and extracted with a dual-stage system in which the first extraction grid was pulsed. They reduced the ratio of the fields (E_1/E_0) in the two extraction regions in order to produce a space-focus plane very close to the source. The reason for this is interesting. While space-focusing converts the initial spatial distribution into a kinetic energy distribution at the space-focus plane, there is an initial kinetic energy distribution in the source itself, which results in expansion of the ion packet (in space and time) as it travels down the drift tube. By shortening the space-focus distance, this problem was minimized, and the space-focus plane served as a suitable origin for the reflectron system.

Their reflectron instrument consisted of two linear drift regions (L_1 and L_2) and the reflectron itself. Ions leaving the source were deflected by an angle a = 2° in order to locate the detector alongside the source region. The Mamyrin reflectron was a two-stage design composed of a retarding region (d_T) and a reflecting region (d_K) in which the direction of ion motion was reversed. In their design, more than 2/3 of the average ion energy (eU_0) was reduced in the short ($d_T = 0.008L$, where $L = L_1 + L_2$) retarding region in which the potential difference between the two grids was $U_T = 0.7U_0$. The ions were reflected in a longer region ($d_K = 0.06L$), where the additional potential difference ($U_K = 0.45U_0$) ensured that the ions turned around before reaching the back of the reflectron.

The performance of the Mamyrin instrument can be appreciated from the mass spectrum of the isotopic cluster of rhenium bromide trimer ions shown in Figure 3.2, in which mass resolution exceeds 1 part in 3000.

The Single-Stage Reflectron

The principles of operation of the reflectron as an energy-focusing device can be more easily understood by considering designs in which only a single retarding/reflecting field is utilized. The single-stage reflectron has been used extensively

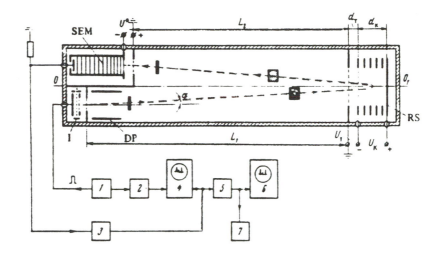

FIGURE 3.1   Diagram of the Mamyrin reflectron showing the ion source (I), deflecting plates (DP), reflecting system (RS), electron multiplier (SEM), rectangular pulse generator (1), delay line (2), wideband amplifier (3), wideband oscillograph (4), stroboscopic attachment (5), low-frequency oscillograph (6), and electronic potentiometer (7). (Reprinted with permission from reference 8).

by Standing's group at Manitoba for focusing the secondary ions formed from surfaces by pulsed bombardment with primary ions in the keV energy range. Figure 3.3 shows a diagram of their pulsed SIMS (secondary ion mass spectrometer) instrument.[9] A resistance network between the entrance grid and a series of lens elements produces a linear voltage drop (constant retarding field) throughout the entire reflectron region. This simpler arrangement is generally referred to as an ion mirror, and very simple relationships exist between its dimensions (and voltages) and the dimensions of the drift lengths.

Energy Focusing in a Single-Stage Reflectron Instrument.   The total flight time of an ion of energy eV in a single-stage reflectron instrument (ignoring for the moment the time spent in the source–extraction regions) is:

$$t = \left(\frac{m}{2eV}\right)^{1/2}\left[L_1 + L_2 + 4d\right]$$ [1]

which shows the same square-root dependence upon mass as linear instruments. The distance d is not the actual length of the reflectron but the distance at which the ions turn around. Referred to as the penetration depth it can be set at some distance

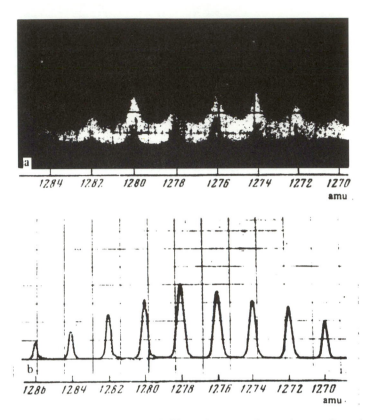

FIGURE 3.2 (a) Oscillographic trace and (b) stroboscopic (boxcar) recording of a portion of the mass spectrum of rhenium bromide showing the isotopic cluster from the trimer. (Reprinted with permission from reference 8).

short of the reflectron depth by setting the voltage at the back of the reflectron to some value greater than the initial accelerating voltage (V) at the source backing plate. The factor of 4 in equation [1] arises from the fact that ions make two passes through the reflectron and (since their average velocities in retarding and accelerating fields are half those in the drift region) spend twice as much time covering the same distance.

The focusing action of the reflectron comes from the fact that the penetration depth varies with ion kinetic energy, while the drift lengths (L_1 and L_2) are constant. In particular, the flight time of an ion with excess kinetic energy (U_0) could be described as:

$$t' = \left(\frac{m}{2(eV + U_0)}\right)^{1/2} \left[L_1 + L_2 + d'\right] \qquad [2]$$

where the reflectron penetration depth has increased:

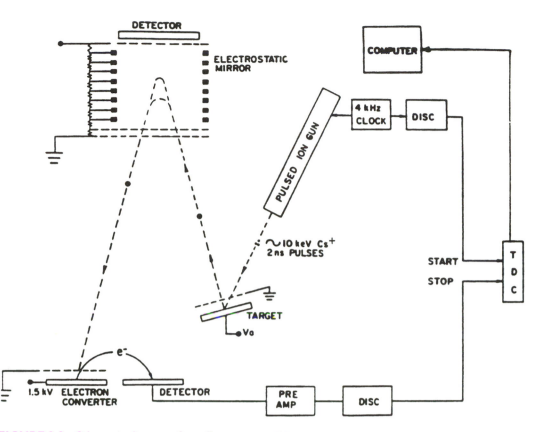

FIGURE 3.3 Schematic diagram of a reflecting time-of-flight mass spectrometer (Manitoba II). (Reprinted with permission from reference 9).

$$d' = \left(\frac{eV + U_0}{eV} \right) d \qquad [3]$$

linearly with the increase in kinetic energy. It should be obvious that the increase in the magnitude of the denominator in equation [2] will shorten the time that an ion with kinetic energy $eV + U_0$ will spend in the drift regions (L_1 and L_2). However, the penetration depth (d) will increase, resulting in longer times in the reflectron and the same total time-of-flight. Thus, within a narrow range, the flight times become independent of the initial kinetic energy U_0. The best resolution is achieved when $L_1 + L_2 = 4d$, that is, when the ions spend equal times in the drift and reflectron regions.[8]

One can easily show that the decrease in time Δt_L spent in the linear regions by an ion with excess kinetic energy will exactly equal the increase in time Δt_R spent in the reflectron when $L_1 + L_2 = 4d$. If $L = L_1 + L_2$, then the difference in time spent in the linear regions will be:

$$\Delta t_L = \left(\frac{m}{2}\right)^{1/2} L \left[\frac{1}{(eV + U_0)^{1/2}} - \frac{1}{(eV)^{1/2}}\right]$$

$$\Delta t_L = \left(\frac{m}{2}\right)^{1/2} L \left[\frac{(eV)^{1/2} - (eV + U_0)^{1/2}}{eV}\right]$$

For small values of U_0 we can expand the term $(eV + U_0)^{1/2}$ to first order:

$$\Delta t_L = \left(\frac{m}{2}\right)^{1/2} L \left[\frac{(eV)^{1/2} - \left((eV)^{1/2} + \dfrac{U_0}{2(eV)^{1/2}}\right)}{eV}\right]$$

$$\Delta t_L = - \left(\frac{m}{2}\right)^{1/2} L \frac{U_0}{2(eV)^{3/2}}$$

Similarly, the increase in time spent in the reflectron will be:

$$\Delta t_R = \left(\frac{m}{2(eV + U_0)}\right)^{1/2} 4 \left(\frac{eV + U_0}{eV}\right) d - \left(\frac{m}{2eV}\right)^{1/2} 4d$$

$$\Delta t_R = \left(\frac{m}{2}\right)^{1/2} 4d \left[\frac{(eV + U_0)^{1/2}}{eV} - \left(\frac{1}{eV}\right)^{1/2}\right]$$

We again expand the term $(eV + U_0)^{1/2}$ to first order:

$$\Delta t_R = \left(\frac{m}{2}\right)^{1/2} 4d \left[\frac{(eV)^{1/2} + \dfrac{U_0}{2(eV)^{1/2}}}{eV} - \frac{(eV)^{1/2}}{eV}\right]$$

$$\Delta t_R = \left(\frac{m}{2}\right)^{1/2} 4d \frac{U_0}{2(eV)^{3/2}}$$

Thus, the difference in time-of-flight is:

$$\Delta t = \Delta t_R + \Delta t_L = \left(\frac{m}{2}\right)^{1/2} \frac{U_0}{2(eV)^{3/2}} [4d - L]$$

and $\Delta t = 0$ when $L = 4d$ (to first order).

Single-stage reflectron instruments are relatively easy to design. For example, if the accelerating voltage is 5 kV, then a reflectron voltage (V_R) of 5100 V would be applied to the back of a reflectron whose total length is $(5100/5000)d$, where d is one-fourth the combined lengths of L_1 and L_2. The drift lengths (L_1 and L_2) can be distributed in any manner, but in most cases L_1 will be approximately equal to L_2 to achieve physical separation of the source and detector at a minimum deflection angle. The reflectron is generally constructed from a series of ring electrodes with inside diameters sufficient to accomodate incoming and outgoing ion trajectories which diverge by the deflection angle. The entrance electrode is generally a grid to prevent lensing effects at the discontinuity (boundary) between the flight tube and the retarding field. The voltage on the entrance grid is the same as that of the flight tube (generally ground potential), and the voltages of all other lens electrodes are provided by a resistance voltage divider connected between ground and V_R.

Analogy Between Single-Stage Reflectron and Space Focusing.   The fact that optimal focusing is achieved when the ions spend an equal time in the reflectron and drift regions should be reminiscent of the single-stage space-focusing problem. The similarity between these two focusing situations is illustrated in Figure 3.4a, where the lengths of L_1 and L_2 have been made equal. Single-stage space-focusing ensures that ions formed over the region $s_0 + \Delta s_0$ are focused at a point ($d = 2s_0$) where the spatial distribution has been converted to a kinetic energy distribution. The inverse process occurs in the retarding phase of reflectron focusing, where the packet of ions formed at the space-focus plane is distributed over a range of penetration depths $d + \Delta d$, where $L_1 = 2d$. In the reacceleration phase, the distribution of penetration depth in the reflectron is spatially refocused to the point $L_2 = 2d$. Thus, the reflectron is essentially a scaled-up version of the space-focusing extraction optics, since it is also intended to achieve wide mass dispersion. In addition, while ions are generally formed in the center of the source, the reflectron utilizes a much larger portion of the acceleration depth.

The single-stage reflectron provides a first-order correction for the kinetic energy distribution. If one carries the analogy with space-focusing further, it should be obvious that dual-stage reflectrons can be designed that push the energy focusing plane well beyond the reflectron depth, and thus enable the use of much smaller reflectrons. In addition, it should also be possible to design second-order reflectrons.

The Dual-Stage Reflectron

The dual-stage reflectron is composed of two linear retarding voltage (constant field) regions, separated by an additional grid (Figure 3.4b). The field strength in the first region is made considerably greater than that in the second region, usually by imposing a larger voltage drop across a shorter distance. Any dual-stage combination in which the retarding field strength is greater in the first stage will (in fact) result in $L_1 + L_2$ being much greater than $4d$ and will enable dual-stage reflectron designs that are considerably smaller with respect to the flight-tube length than single-stage designs.

If space were the only consideration, such first-order designs would indeed be worthwhile. However, because the original Mamyrin reflectron provided second-order energy correction, dual-stage reflectrons have always taken advantage of this superior focusing capability. Dual-stage reflectrons can be easily designed. The general

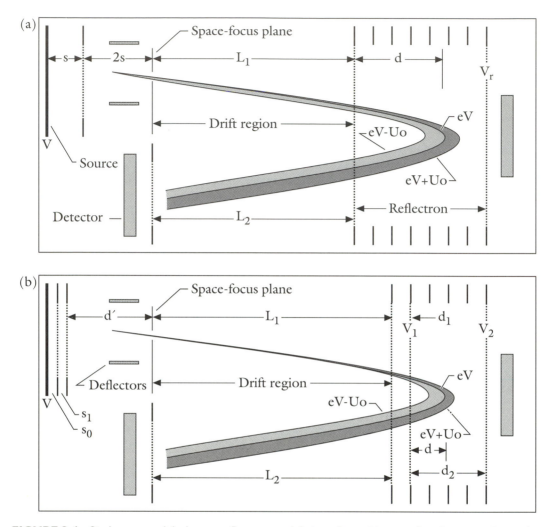

FIGURE 3.4   Single-stage and dual-stage reflectrons and their analogy with space-focusing extraction optics: (a) single-stage ion extraction and a single-stage ion mirror, and (b) dual-stage extraction and a dual-stage reflectron.

rule of thumb is that the ions should lose about two-thirds of their kinetic energy in the first one-tenth of the reflectron depth. The flight-tube length (L_1 or L_2) will then be about 8 times the length of the reflectron. The voltages on the reflectron can then (of course) be finely adjusted to produce maximum resolution.

In the Mamyrin reflectron the voltage drop across the first stage (d_1 = 10 mm) was $0.7V$, where V = the acceleration voltage, while the voltage drop across the second stage (d_2 = 100 mm) was $0.45V$. While their instrument used an ion source at ground and pulsed ion extraction, dual-stage reflectron instruments based upon their dimensions that use high-voltage backing plates would place the voltage at the back of the reflectron at $(0.7 + 0.45)V = 1.15V$ and the turn-around region would be located at two-thirds the depth of the second region.

If one recalls the equation that was used to determine the space-focus plane for a dual-stage extraction system:

$$d = 2\sigma^{3/2}\left[\frac{1}{s_0^{1/2}} - \frac{2s_1}{s_0^{1/2}(\sigma^{1/2} + s_0^{1/2})^2}\right]$$

where $\sigma = s_0 + (E_1/E_0)s_1$, then an analogous expression can be written for the retarding phase of the reflectron:

$$L_1 = 2\delta^{3/2}\left[\frac{1}{d_2^{1/2}} - \frac{2d_1}{d_2^{1/2}(\delta^{1/2} + d_2^{1/2})^2}\right]$$

where $\delta = d_2 + (\Delta V_1 d_2/\Delta V_2 d_1)d_1$, ΔV_1 is the voltage drop across the first stage d_1, and ΔV_2 is the voltage drop across the second stage d_2.

In the Mamyrin reflectron, $\Delta V_1 = 0.7\,V$, $\Delta V_2 = 0.45\,V$, $d_1 = 10$ mm, and $d_2 = 100$ mm, so that $\delta = 100 + (0.7/0.45)(100/10)10 = 255$, and:

$$L_1 = 2(255)^{3/2}\left[\frac{1}{(100)^{1/2}} - \frac{2(10)}{(100)^{1/2}(255^{1/2} + 100^{1/2})^2}\right]$$

$$L_1 = 2(4096)[0.1 - 0.00296] = 795\ mm$$

or about 8 times the length of d_2. The actual drift lengths in the Mamyrin instrument were $L_1 = 820$ mm and $L_2 = 780$ mm.

Although we have used the first-order equations to calculate the focal point, the Mamyrin reflectron was designed for second-order focusing. A similar reflectron has been designed by LeBeyec and co-workers[10] that can accomodate a wide range of drift-tube lengths, and provides an excellent starting point for designing a high-resolution time-of-flight mass spectrometer. As shown in Figure 3.5, it consists of 11 ring-type lenses and 3 grids, connected by the resistors R_1 to R_{13}. The distance between each of the grid and lens elements is 1 cm, so that the overall length of the reflectron is 13 cm. The resistors R_1 to R_{12} are all 2 MΩ, while R_{13} is variable to permit reduction of one-half to two-thirds of the ion kinetic energy within the first one-twelfth of the reflectron length. By adjusting the voltage U_{G3} at the back of the reflectron and the voltage U_{G2} on the second grid (using R_{13}), the reflectron can be used with accelerating voltages as high as 20 kV and can provide second-order focusing for drift lengths ($L_1 + L_2$) of 1.2 to 2 m. In addition, their dimensions can be scaled up or down to provide focusing for instruments with larger or smaller drift lengths.

The second-order, dual-stage reflectron provides energy focusing over a broader range of kinetic energies than the single-stage reflectron. The performance of the dual-stage reflectron instrument reported by LeBeyec[10] is shown in Figure 3.6. Flight times of ions with $\Delta U/U_0$ up to $0.025\,V$ are unchanged, so that a 5-keV instrument

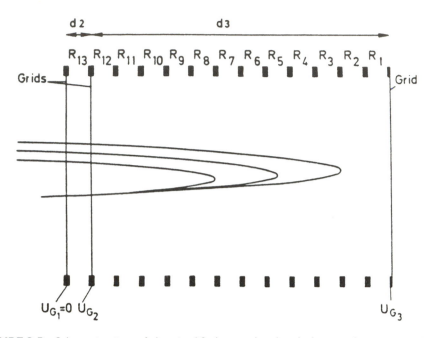

FIGURE 3.5 Schematic view of the simplified second-order, dual-stage electrostatic mirror of LeBeyec and co-workers, showing the trajectories of ions having kinetic energies of 0.9eV, eV, and 1.1eV, where V = the accelerating voltage. (Reprinted with permission from reference 10).

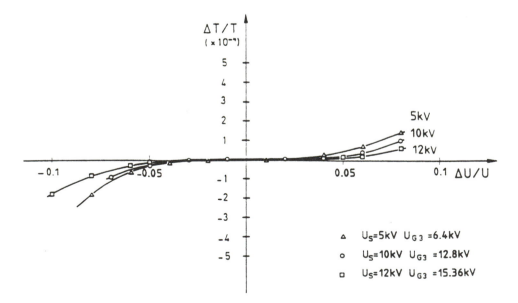

FIGURE 3.6 Experimental relative variations of $\Delta T/T$ of the peak centroid of Cs_2I^+ formed by PDMS as a function of variations $\Delta U/U$ of the accelerating voltage. U_{G3} is the total voltage applied to the reflectron. (Reprinted with permission from reference 10).

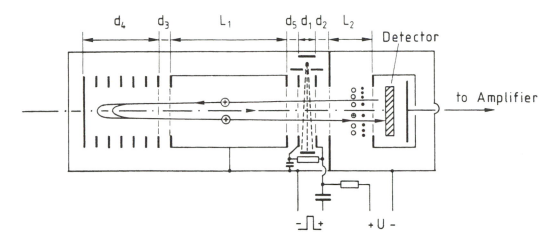

FIGURE 3.7   A linear reflectron with transparent ion source. (Reprinted with permission from reference 12).

will tolerate energy spreads of ±125 eV. Nevertheless, there are some disadvantages to dual-stage reflectrons, resulting from the addition of a second grid. In addition to the normal transmission losses resulting from ions striking the grid wires, the close spacing of two grids having very different voltages results in high local electrical fields in the vicinity of the grid wires that produces additional ion deflection and scattering.

Coaxial Reflectrons

The angle between the incoming and outgoing trajectories of the reflectron analyzer is necessary to permit physical separation of the source and detector. Figure 3.4 shows that the angle, although small (1.4° to 3.0°), causes considerable broadening of the outgoing ion beam because of the different penetration depths of ions with different kinetic energies.

An alternative is to place the source, reflectron, and detector on the same axis. An early example of this configuration was published in 1979 by the Mamyrin group[11,12] and is shown in Figure 3.7. In this instrument the detector is located behind a *transparent* ion source. The ion source in this case is an electron-beam source, bounded by two (backing and extraction) grids, which can be pulsed to eject the ions toward a dual-stage reflectron. By the time that the ions return to the source region, these two grids are placed at the potential of the flight tube, so that the source becomes transparent to the ions, and their trajectories are undisturbed as they move toward the detector. This approach works well with the pulsed EI source, but would be more difficult to implement on a laser desorption source in which ions are formed (usually) on the backing plate. In addition, if very broad mass ranges are to be recorded, heavy ions would continue to be extracted from the source at the time that light ions have returned.

Dual-channelplate detectors that have a small, centered hole through which the incoming ion beam can be transmitted are now available commercially, and have been used in a number of coaxial designs in which the detector is placed between the ion source and the reflectron. Della Negra and Le Beyec[13] reported such an instrument in 1984. This instrument was designed for the analysis of metastable product ions using the *correlated reflex technique* that is discussed in detail in Chapter 8. The instrument used by Tanaka and co-workers[14] in their first report on matrix-assisted laser desorption also incorporated a coaxial reflectron (Figure 3.8). The reflectron in this instrument was also a quadratic reflectron, designed to produce time-focusing independent of kinetic energy. This type of reflectron is described in more detail below in the section on ideal reflectrons.

Correction for the Extraction Region

The most common approach to time-of-flight design minimizes temporal distributions by using short-pulsed-laser or particle beams, and reduces or compensates for spatial distributions by desorbing ions from an equipotential surface or by dual-stage extraction, respectively, thus leaving only kinetic energy distributions to be corrected by a reflectron. The dimensions of the drift (L_1 and L_2) and reflectron regions are then chosen to provide focusing of the kinetic energy distribution of the ion packet as it appears at the entrance to the drift region or (in the case of dual-stage extraction) at the space-focus plane. This approach to reflectron design assumes that the time spent in the source–extraction region is negligible, particularly when very short, high-field extraction regions are used. However, the effect of initial kinetic energy distribution in the source is to produce a temporal (as well as kinetic energy) distribution at the entrance to the flight tube (or at the space-focus plane when there are both initial kinetic energy and spatial distributions), which cannot be compensated by the reflectron. Short and Todd[15] have noted that in many cases the detrimental effects of the source–extraction region may override the benefits of second-order focusing provided by dual-stage reflectrons, unless particularly long flight tubes are utilized to minimize the effects of the temporal distribution. Thus, these investigators have developed reflectron designs which utilize an additional high-field retarding region toward the back of the reflectron to compensate for the effects of the extraction region. Their motivation arose from the need to design an instrument with a broad extraction region that could accomodate a rastered primary ion beam for ion imaging. In addition, their instrument utilized low ion-accelerating voltages, rather than an extended flight-tube length, to increase the time that the ions would spend in the flight tube.

Figure 3.9 shows the short and Todd modification of a two-stage reflectron to a three-stage reflectron, which provides second-order energy compensation for the flight times in both the extraction and drift regions. Their instrument utilizes an ion-accelerating voltage of 200 V and a reflectron voltage of 224 V to focus ions formed in a source region with a broad extraction length (2.5 cm). The total drift length is 58.4 cm, and the total reflectron length is 5.75 cm, giving an overall length for the instrument of about 43 cm. Figure 3.10 shows the focusing obtained as the ratio of the fields in the second and third regions varies from the case where the reflectron acts as a two-stage reflectron through its optimal focusing condition.[15]

An identical three-stage reflectron design has been reported by Vestal and Nelson[16,17], but was implemented on a high accelerating voltage instrument with a short extraction region and long drift region. Thus, the improvements in mass resolution (in their case) were less dramatic, since the resolution was considerably higher without compensation than that observed in the Short and Todd instrument.

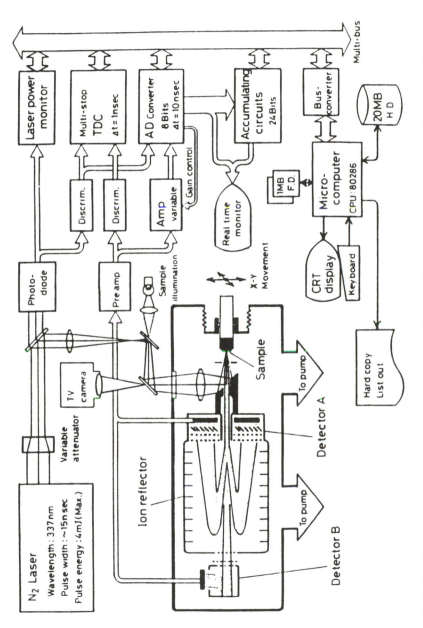

FIGURE 3.8   Schematic diagram of the coaxial reflectron time-of-flight mass spectrometer used by Tanaka et al. (Reprinted with permission from reference 14).

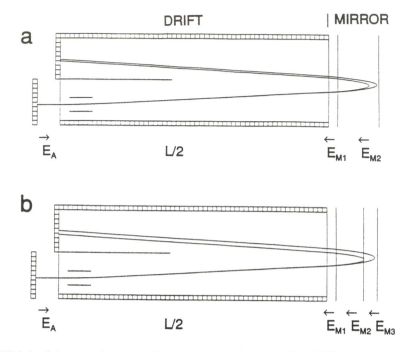

FIGURE 3.9   Schematic diagram and simulated ion trajectories for a TOF mass spectrometer (a) with a conventional two-stage mirror, and (b) with the three-stage mirror designed by Short and Todd. (Reprinted with permission from reference 15).

Overall, such reflectrons may prove to be most beneficial for the low voltage, long extraction region configurations described by Short and Todd.[15] For instruments in which the time spent in the source–extraction region has been considerably minimized, the resolution benefits may not outweigh the transmission losses due to scattering of low velocity ions that turn around in the vicinity of the grid.

Ideal Reflectrons

The reflectrons described thus far (and most commonly employed in time-of-flight instruments) use one or more linear voltage (constant field) regions for retarding, reflecting, and reaccelerating ions of different kinetic energy. In the single-stage reflectron the voltage on each lens element is $V_x = ax$, where x is the distance from the entrance to the reflectron, and the slope $a = V/d$ is chosen so that the acceleration voltage V is reached at $d = (L_1 + L_2)/4$. The dual-stage reflectron has two linear voltage regions with different slopes.

The ideal reflectron would provide kinetic energy correction to infinite order. One can easily show that this would occur if the reflectron was designed such that $V_x = ax^2$. In this case, the flight time of an ion in the reflectron can be calculated:

$$t = 2\int_0^{x_{max}} \frac{dx}{\left[\frac{2e}{m}\left(V_x - ax^2\right)\right]^{1/2}} = \frac{\pi}{\left(\frac{2ea}{m}\right)^{1/2}} \qquad [4]$$

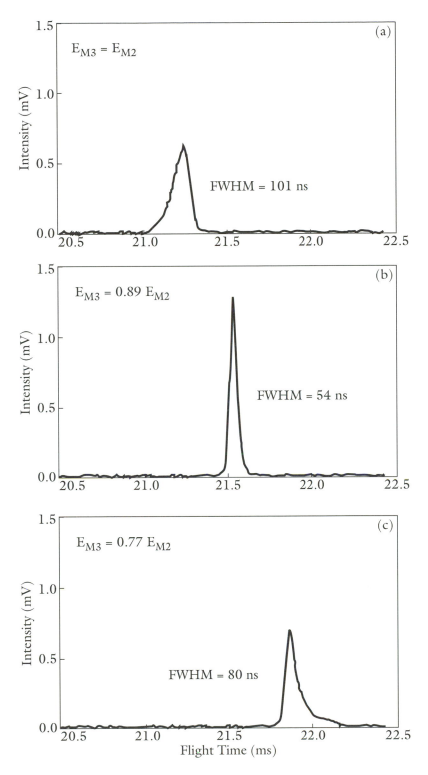

FIGURE 3.10 Time-of-flight mass spectra for Na^+ ions as a function of E_{M3}/E_{M2} ratio for (a) conventional two-stage configuration, (b) optimal temporal focusing, and (c) overcompensation. (Reprinted with permission from reference 15).

which results in a mass scale:

$$t = \pi \left(\frac{m}{2ea} \right)^{1/2} = km^{1/2} \qquad\qquad [5]$$

which is square root in mass, *and completely independent of kinetic energy.*[12] In such a reflectron arrangement, there are no linear drift regions. The entrance of the reflectron (and the detector) would be located at the end of the source–extraction region (but would not, of course, compensate for time spent in that region) or at the space-focus plane.

Quadratic reflectrons are difficult to implement experimentally. When constructed from a lens stack so that the voltage on each lens (x) is $V_x = ax^2$, the voltage along the central axis will not follow this law, and will result in radial components of the electrical field that will cause divergence of the ion trajectories for ions that are slightly off-axis. These problems can be illustrated by comparing the SIMION (simulated ion trajectory) for a single-stage (Figure 3.11a), dual-stage (Figure 3.11b), and quadratic reflectron (Figure 3.11c). Nonetheless, in 1986 Yoshida and co-workers at the Shimadzu Corporation (Kyoto) obtained a patent[18] for a coaxial reflectron (Figure 3.12) using a large number (100) of closely spaced, small-inner-diameter (40 mm) lens elements to achieve conformity of the axial voltages with those imposed on the lens elements. This instrument was used in work described in their first report[14] of matrix-assisted laser desorption of proteins (see Chapter 6).

In 1993, Davis and Evans filed a patent application[19] for a tandem sector–TOF mass spectrometer that incorporated a novel reflectron designed to overcome the problems associated with stacked lens elements. The reflectron was constructed in a monopole geometry as shown in Figure 3.13, where the voltage along the x-axis (between a grounded V-shaped lens and a cyclindrical lens carrying the reflectron voltage) follows the law $V = ax^2$. The ions enter and exit through a slit in the V-lens. While this arrangement does indeed deliver the promised quadratic retarding voltage, the curved equipotential lines suggest that it will again defocus ions entering off-axis. The problem may be further exacerbated by the lensing effects that may be introduced when the ion entrance–exit slit is present at the apex of the V-lens.

Gridless Reflectrons

Because scattering occurs when ions pass through a grid and can result in loss of ion transmission, there has been considerable interest in developing gridless reflectrons. A design reported by Wollnik et al.[20] is shown in Figure 3.14. To date, such reflectrons are not widely used, perhaps because they require tight focusing of the ion beam close to the reflectron axis, since (as shown) the equipotential lines are not parallel and can result in divergence of the ion trajectories for off-axis ions. The SIMION plot in Figure 3.11d illustrates the similarities of the Wollnik reflectron with a dual-stage reflectron, where the absence of grids provides a smooth transition at the entrance of the reflectron and at the boundary of the regions with different retarding electrical fields.

Despite the considerable appeal for utilizing gridless reflectrons to reduce losses in ion transmission due to scattering, designs which utilize grids to separate and

(a)

(b)

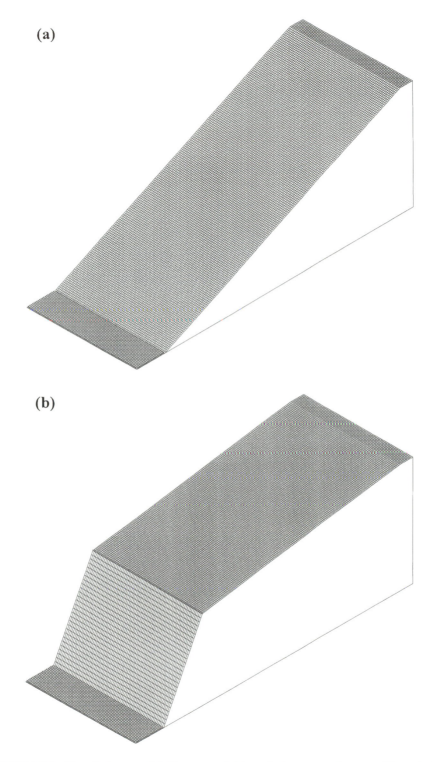

FIGURE 3.11 Electrical potential diagrams of (a) a single-stage reflectron, (b) a dual-stage reflectron, (c) a quadratic reflectron, and (d) the gridless reflectron described by Wollnik and co-workers,[20] generated by SIMION.

(c)

(d)

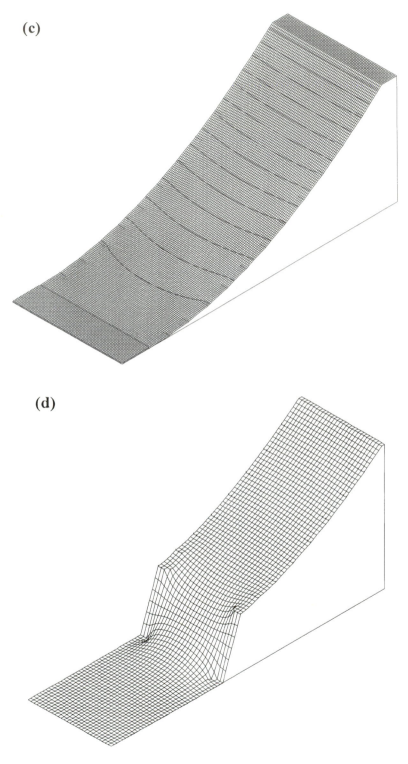

FIGURE 3.11 (continued)

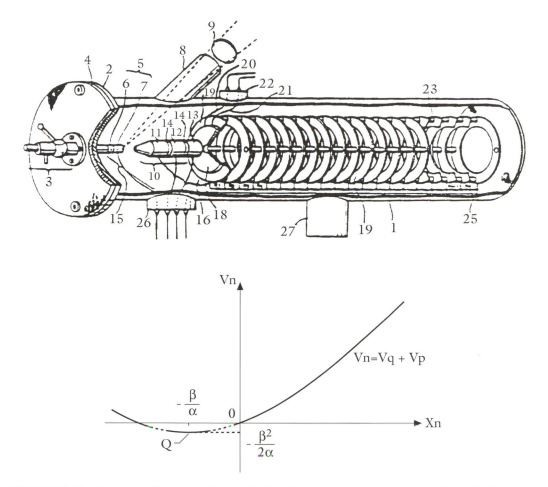

FIGURE 3.12 Schematic diagram of the quadratic reflectron mass spectrometer of Yoshida. (Reprinted from reference 18).

define regions with different retarding fields continue to be dominant. Such reflectrons are considerably easier to design, and (because the field lines are parallel) can accommodate a wider range of incoming trajectories.

Electrostatic Energy Analyzers

Both time-of-flight and magnetic-sector mass spectrometers require acceleration of ions to a constant energy ($mv^2/2 = eV$). In both types of instruments resolution is degraded by the initial kinetic energy distribution, but in magnetic mass spectrometers the problem has been successfully addressed by the addition of an electrostatic energy analyzer (ESA), which reduces the energy spread and acts as a narrow bandpass filter for ions within a narrow energy range, resulting in mass resolutions greater than 1/100,000 on high-performance instruments. While it is tempting to assume that electric sectors can provide the same benefits for time-of-flight mass spectrometers, it

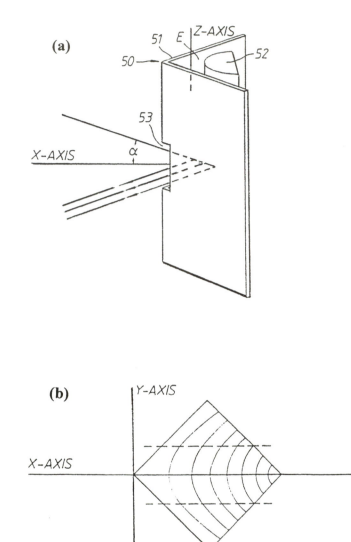

FIGURE 3.13 The monopole reflectron of Davis and Evans: (a) construction and (b) diagram showing the equipotential lines. (Reprinted from reference 19).

is important to remember that the task is not to reduce the kinetic energy spread, but to insure that ions with different kinetic energies (but the same mass) arrive at a detector at the same time. The electric sectors used in double-focusing instruments, in fact, perform quite the reverse function. A narrow ion packet formed in the ion source (for example, from a pulsed laser) will be energy-focused in the ESA by slowing down the more energetic ions and speeding up the less energetic ions, resulting in a considerably wider ion packet after the ESA.

Nonetheless, electrostatic energy analyzers have played a significant role in time-of-flight design. One approach, exemplified by the ETOF mass spectrometer developed at

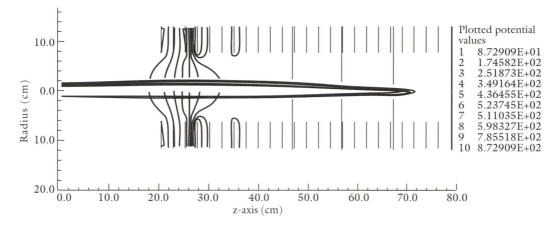

FIGURE 3.14 Diagram of the gridless reflectron described by Wollnik et al.,[20] showing the equipotential lines. (Reprinted with permission from reference 20).

Michigan State University,[21] forms ions continuously, and gates the ions *after* the ESA to produce a narrow time packet. Gating is usually accomplished by passing the ions through a pair of deflector plates, with one deflector plate at ground potential and the other at some positive voltage, and applying a square pulse to the second deflector to return it momentarily to ground (Figure 3.15a). The width of the ion packet in this case is the width of the square pulse. Shorter ion packets can be produced by rapidly reversing the voltages on the two deflectors (Figure 3.15b). When the deflected ion beam is swept across a narrow slit, very short ion packets can be produced. As described in Chapter 5, the SIMS instruments of Benninghoven[22] and Standing[23] used such schemes for producing narrow (1 ns) primary ion-beam packets. Alternatively, gating can be accomplished by rapid reversal of the polarities of a series of wires of alternating polarity (Figure 3.15c).[24]

A second approach is to design electrostatic devices that focus ions with respect to both energy and time, referred to generally as isochronous energy focusing. An isochronous electrostatic energy analyzer, designed in 1972 by Poschenroeder,[25] was utilized by Benninghoven and co-workers[26] in their static SIMS time-of-flight instrument (Figure 3.16). Isochronous energy focusing can also be achieved by using multiple electric sectors. Sakurai et al.[27] utilized a symmetric arrangment of four electrostatic sectors (shown in Figure 3.17) that achieved simultaneous third-order energy and space focusing. Each of the electric sectors in this instrument had a radius of 5 cm and focused ions through an angle of 269°. This provided a compact arrangement that nonetheless had an overall flight length of 1.7 m. Most interesting is a configuration developed by Schueler et al.[28] using three electric sectors. Shown in Figure 3.18, this instrument provides isochronous energy focusing while preserving spatial information, and is thus used for ion imaging.

Electric sectors have not enjoyed anywhere near the same popularity as reflectrons in time-of-flight design. Isochronous systems, particularly those involving multiple sectors, are more difficult to design. In addition, reflectron instruments provide an important opportunity for recording the structural information provide by post-source decay, which is now being utilized for the amino acid sequencing of peptides.

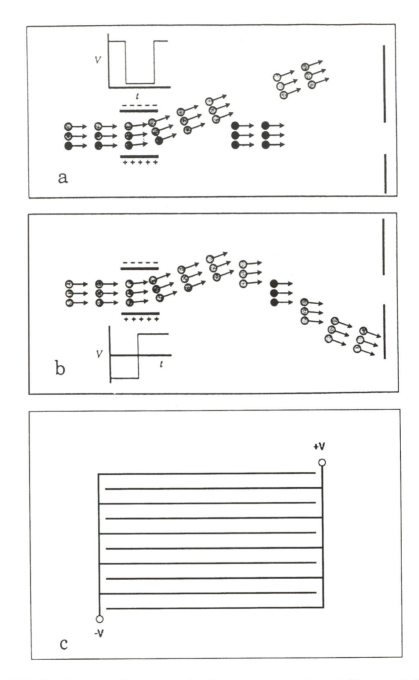

FIGURE 3.15 Formation of ion packets by (a) a square wave pulse, and (b) reversal of voltages on deflection plates. A series of wires of alternating polarity that make up a Bradbury-Nielsen gate is shown in (C).[24]

Linear vs. Reflectron Instruments

Reports by Chait and co-workers[29,30] and Demirev et al.[31] demonstrated that metastable fragmentation of biological molecules (with molecular weights greater than 1 kDa) occurs in a time frame that encompasses their flight times in the drift

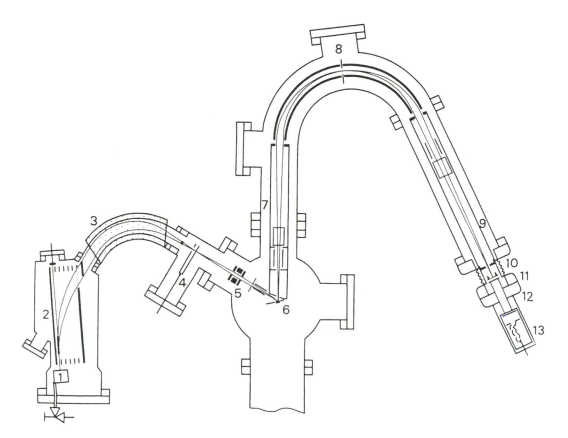

FIGURE 3.16 Pulsed SIMS time-of-flight mass spectrometer of Benninghoven et al., using an isochronous electrostatic energy analyzer: (1) primary ion source, (2) deflector, (3) bunching magnet, (4) start pulse multiplier, (5) lens, (6) target, (7 and 9) linear drift regions, (8) 163° toroidal condenser, (10) post-acceleration gap, (11) channelplate, (12) scintillator and light guide, and (13) photomultiplier. (Reprinted with permission from reference 22).

region. Fragmentation in the drift region produces product ions with the same velocity as their precursors; and if no additional electrical fields are encountered (e.g., reflectrons or electric sectors), they will arrive at the detector with the same flight times as their precursor ions. Thus, if the objective is to determine the molecular weight of a large biomolecule (or to utilize any number of molecular weight strategies, such as ladder sequencing), the linear TOF mass spectrometer provides the highest sensitivity. For this reason, linear time-of-flight mass spectra still comprise by far the majority of spectra in the literature when TOF instruments are used for biological structural applications.

Reflectron (RTOF) mass spectrometers are generally designed so that they can record either linear or reflected time-of-flight mass spectra. In addition to improved mass resolution, reflectron instruments provide the opportunity to obtain structural information from the metastable fragmentation that occurs in the flight tube. In one mode of operation, reflected product ions are correlated with neutral species recorded at the back of the reflectron. In another mode, molecular ions are mass-selected by timed gating in the first drift region (L_1), and the product ions recorded in reflected

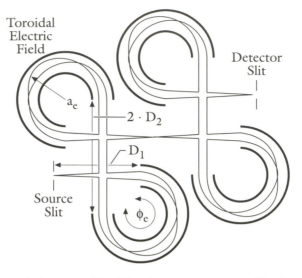

FIGURE 3.17   Ion optical system of the Sakurai mass spectrometer. (Reprinted with permission from reference 27).

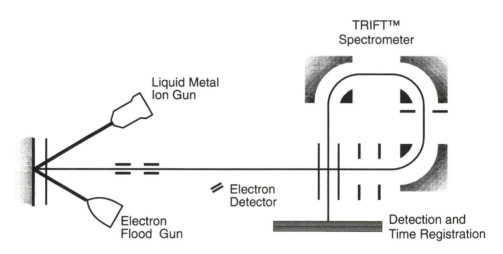

FIGURE 3.18   Diagram of the isochronous and spatial-focusing time-of-flight mass spectrometer described by Scheuler et al. (Reprinted with permission from reference 28).

spectra, either by scanning the reflectron voltage or by utilizing reflectron designs that provide focusing over a wide range of ion kinetic energy. These post-source decay methods provide perhaps the most important role for reflectron analyzers and are described in detail in Chapter 8.

References

1. Wiley, W.C.; McLaren, I.H., *Rev. Sci. Instr.* **26** (1955) 1150–1157.
2. Marable, N.L.; Sanzone, G., *Int. J. Mass Spectrom. Ion and Phys.* **13** (1974) 185–194.
3. Browder, J.A.; Miller, R.L.; Thomas, W.A.; Sanzone, G., *Int. J. Mass Spectrom. Ion and Phys.* **37** (1981) 99.
4. Muga, M.L., *Anal. Instrum.* **16** (1987) 31.
5. Yefchak, G.E.; Enke, C.G.; Holland, J.F., *Int. J. Mass Spectrom. Ion Processes* **87** (1989) 313.
6. Kinsel, G.R.; Johnston, M.V., *Int. J. Mass Spectrom. Ion Processes* **91** (1989) 157.
7. Kinsel, G.R.; Grundwürmer, J.M.; Grotemeyer, J., *J. Am. Soc. Mass Spectrom.* **4** (1993) 2.
8. Mamyrin, B.A.; Karataev, V.I.; Shmikk, D.V.; Zagulin, V.A., *Sov. Phys. JETP* **37** (1973) 45.
9. Tang, X.; Beavis, R.; Ens, W.; LaFortune, F.; Schueler, B.; Standing, K.G., *Int. J. Mass Spectrom. Ion Processes* **85** (1988) 43–67.
10. Brunelle, A.; Della-Negra, S.; Depauw, J.; Joret, H.; Le Beyec, Y., *Rapid Commun. Mass Spectrom.* **5** (1991) 40–43.
11. Mamyrin, B.A.; Shmikk, D.V., *Sov. Phys.* **49** (1979) 762.
12. Mamyrin, B.A., *Int. J. Mass Spectrom. Ion Processes* **131** (1994) 1–19.
13. Della Negra, S.; Le Beyec, Y., *Int. J. Mass Spectrom. Ion Processes* **61** (1984) 21.
14. Tanaka, K.; Ido, Y.; Akita, S.; Yoshida, Y.; Yoshida, T., *Rapid Commun. Mass Spectrom.* **2** (1988) 151.
15. Short, R.T.; Todd, P.J., *J. Am. Soc. Mass Spectrom.* **5** (1994) 779–787.
16. Vestal, M.L.; Nelson, R.W., *Proceedings of the 40th ASMS Conference on Mass Spectrometry and Allied Topics*, Washington, DC May 1992, p. 350.
17. Vestal, M.L., U.S. Patent 5,160,840 (1992).
18. Yoshida, Y., U.S. Patent 4,625,112 (1984).
19. Davis, S.C.; Evans, S., European Patent Application EP 0 551 999 A1 (1993).
20. Grix, R.; Kutscher, R.; Li, G.; Grüner, U.; Wollnik, H., *Rapid Commun. Mass Spectrom.* **2** (1988) 83–85.
21. Pinkston, J.D.; Robb, M.; Watson, J.T.; Allison, J., *Rev. Sci. Instr.* **57** (1986) 583.
22. Steffens, P.; Niehuis, E.; Friese, T.; Benninghoven, A., in *Ion Formation from Organic Solids*, Benninghoven, A., Ed.; Springer-Verlag, Berlin (1983) pp. 111–117.
23. Chait, B.T.; Standing, K.G., *Int. J. Mass Spectrom. Ion and Phys.* **40** (1981) 185.
24. Bradbury, N.E.; Nielsen, R.A., *Phys. Rev.* **49** (1936) 388–393.
25. Poschenroeder, W.P., *Int. J. Mass Spectrom. Ion and Phys.* **9** (1972) 357.
26. Steffins, P.; Niehuis, E.; Friese, T.; Greifendorf, D.; Benninghoven, A., *Vac. Sci. Technol.* **A3** (1985) 1322.
27. Sakurai, T.; Fujita, Y.; Matsuo, T.; Matsuda, H.; Katakuse, I.; Miseki, K., *Int. J. Mass Spectrom. Ion Processes* **66** (1985) 283–290.
28. Schueler, B., *Microsc. Microanal. Microstruct.* **3** (1992) 1–21.
29. Chait, B., *Int. J. Mass Spectrom. Ion and Phys.* **53** (1983) 227-.
30. Chait, B.; Field, F.H., *Int. J. Mass Spectrom. Ion Processes* **65** (1985) 169-.
31. Demirev, P.; Olthoff, J.K.; Fenselau, C.; Cotter, R.J., *Anal. Chem.* **59** (1987) 1951-.

4

Plasma Desorption Mass Spectrometry

In 1973, Ron Macfarlane and co-workers at the Texas A&M University Cyclotron Laboratory used energetic fission fragments from a radioactive californium source to bombard a sample of sodium acetate, and recorded the time-of-flight mass spectrum shown in Figure 4.1.[1,2] As indicated on the figure, signed by Macfarlane, Torgerson, and Skowronski, they considered this an historical event, since they quickly went on to show that fission fragments from ^{252}Cf could be used to produce molecular ions from compounds that were considered intractable (i.e., nonvolatile and thermally unstable) by other ionization techniques.[3,4] Their vision was certainly correct, since it was subsequently shown that the technique could be used to desorb and mass analyze intact molecular ions of peptides and small proteins with molecular weights up to 35 kDa.[5]

The transuranium element californium was first observed in the 60-inch cyclotron at Berkeley in 1950, where bombardment of ^{242}Cm with 35-MeV helium ions produced ^{249}Cf.[6] Heavier isotopes of this element are produced by intense neutron bombardment of ^{249}Cf. Thus, ^{252}Cf was detected in the first thermonuclear (hydrogen) bomb test conducted on the Eniwetok atoll in 1952. Californium-252 has a decay half-life of 2.65 years, resulting primarily (97%) in the emission of alpha particles. More interestingly, ^{252}Cf undergoes spontaneous fission (3%), emitting two multiply charged fission fragments simultaneously and in nearly opposite directions. A typical decay would involve the simultaneous emission of ^{106}Tc and ^{142}Ba with energies of 104 and 79 MeV, respectively.[4] Additionally, ^{252}Cf is a strong neutron emitter, with 1 µg of ^{252}Cf providing a neutron flux of about 10^6 neutrons/s. While this is considerably smaller than the neutron flux available in a nuclear reactor, ^{252}Cf has been used for a number of years as a neutron source for elemental analysis by neutron activation.

The technique introduced by the Macfarlane group was originally referred to as Coulomb ionization since the process of excitation was similar to the Coulomb excitation of nuclei by fast heavy ions.[2] The term fission-fragment induced desorption (FFID) was used by Field and Chait[7] to describe the technique without implying a particular mechanism. Other investigators, including Bo Sundqvist at the Tandem

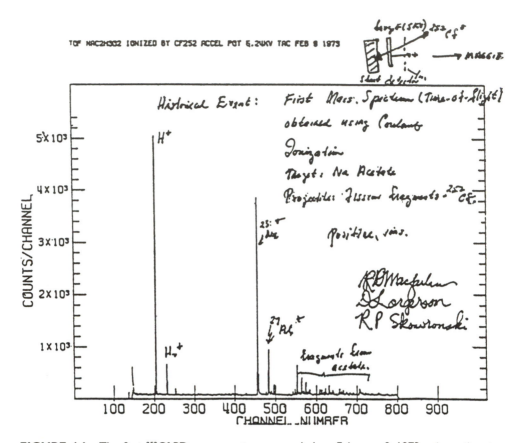

FIGURE 4.1   The first ^{252}Cf-PD mass spectrum recorded on February 9, 1973, using a time-to-amplitude converter and pulse-height analyzer.

Accelerator Laboratory in Uppsala,[8] Karl Wien at Darmstadt,[9] and Yvon LeBeyec at Orsay,[10] produced similar results using heavy particles from linear accelerators. Since these did not involve fission fragments, they suggested heavy-ion induced desorption (HIID) to describe the technique generally and to distinguish it from those techniques, primarily secondary ion mass spectrometry (SIMS), that utilized light ions with energies in the keV range. However, the term plasma desorption mass spectrometry (PDMS) suggested early on by MacFarlane[4] is generally accepted for this technique.

Molecular biologists, biochemists, and immunologists were envisioned as the major user groups for this technique. However, with the exception of a californium-based instrument constructed at the Rockefeller Institute[7] and an instrument at the National Institutes of Health[11] (constructed by the Texas A&M laboratory), PDMS was carried out primarily in nuclear physics laboratories for nearly 10 years. It was not until 1984 that commercial instruments utilizing ^{252}Cf sources became available from BIO-ION Nordic (Uppsala, Sweden). Their first instrument, the BIN-10K, was installed in laboratories at Odense University (Sweden), The Johns Hopkins University (Baltimore, MD), and the Mayo Clinic (Rochester, MN). A later model, the BIO-ION 20, reached worldwide sales of more than 50 instruments by 1993, when BIO-ION announced that it would no longer manufacture plasma desorption instruments.

The Plasma Desorption Mass Spectrometer

The basic elements of a ^{252}Cf-PDMS instrument are shown in Figure 4.2. The 10-μCi ^{252}Cf source produces pairs of fission fragments that are emitted in opposite directions. One of these fission fragments penetrates a thin metal foil toward the back of the spectrometer, which releases a burst of secondary electrons that are detected and amplified by a dual channelplate *start* detector. Alpha particles emitted from the source also produce secondary electrons at the *conversion* foil, but are much fewer in number since the kinetic energy (around 4 MeV) is considerably smaller than that of the heavy fission particles. Because the amplified pulse output from the multiplier is also considerably smaller for α-particles, fission events can be distinguished from α-particle emission using a constant fraction discriminator (CFD), which passes only those pulses due to fission events to the time-to-digital converter (TDC) to initiate a timing cycle. The second fission fragment penetrates a thin aluminum foil holding the sample. Ions desorbed from the sample are accelerated to energies of 10 to 20 keV, pass through the drift tube, and are detected by a *stop* detector. The TDC then determines their flight times by comparing the start and stop times. The plasma desorption mass spectrum is produced by recording and accumulating the timing results from many such cycles. For a 10-μCi source, there will be from 1200 to 1500 such cycles each second depending upon the instrument geometry. If one (typically) accumulates the results from 10^6 such events, the mass spectrum is obtained in about 12 minutes.

Figure 4.3 shows the mass spectrum of insulin obtained on a BIN-10K mass spectrometer, with a flight-tube length of 15 cm. Typical mass resolution on that instrument is 500, so that isotopic contributions to the molecular-ion peak are not resolved. At the same time, mass measurement accuracy is considerably better, ranging from about 0.01% (below 5000 amu) to 0.1% at higher masses.

Coincidence Instruments

In the 1950s the energies of gamma rays from radioactive nuclides were measured using scintillation–photomultiplier detectors, which produced an output pulse whose amplitude was proportional to the gamma-ray energy. This output pulse was then passed to a single channel analyzer, which consisted of low- and high-level discriminators to record only those pulses falling within a preset pulse amplitude. This was a cumbersome technique, since in order to record the energy spectrum it was necessary to manually step the single-channel analyzer to each energy (pulse amplitude) window, and to count pulses for a preset time. In later years, scintillation detectors were replaced with more efficient germanium–silicon (GeLi) and silicon–lithium (SiLi) detectors, and these single channel pulse height analyzers (PHA) were replaced with multichannel analyzers (MCA), which enabled recording of the entire energy spectrum without scanning.

The Time-to-Pulse-Height Converter (TPHC).   To adapt such schemes to time-of-flight mass spectrometers, it was only necessary to convert timing information to pulse-height information. This was accomplished using the time-to-pulse height converter (TPHC). One of the earliest time-of-flight instruments to utilize this scheme was a mass spectrometer developed in the early 1960s at the Johnston

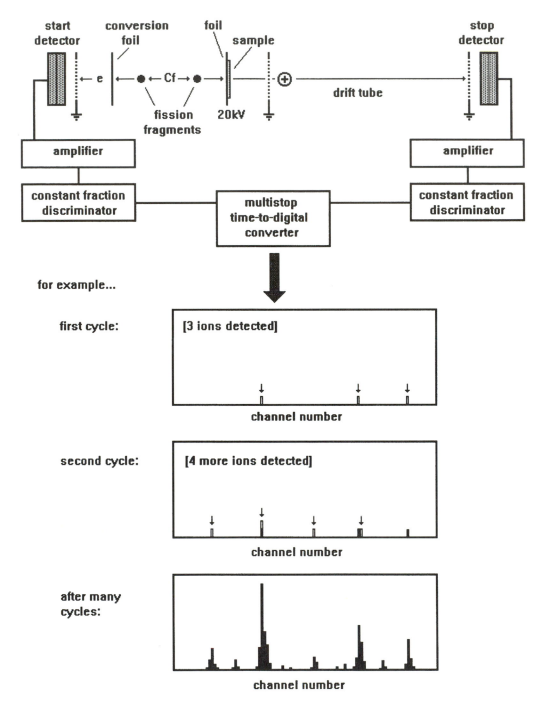

FIGURE 4.2 Schematic diagram of a plasma desorption mass spectrometer. Because the TDC is a multistop device, each ionization and recording cycle produces time-interval measurements for several ions. In the example shown, 3 and 4 ions are recorded in the first and second cycles, respectively.

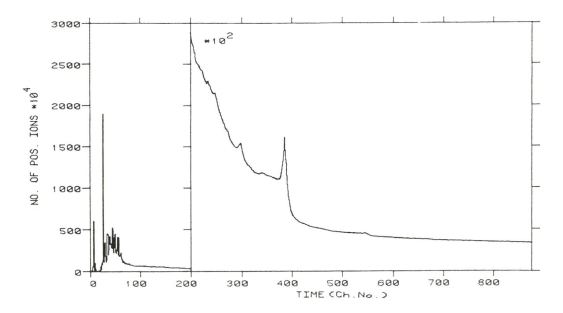

FIGURE 4.3 PDMS spectrum of insulin. Note in the low mass portion of the spectrum that there is a broad peak resulting from the fission fragments themselves, whose velocities $(2eV/m)^{1/2}$ are approximately 1 MeV/nucleon.

Laboratories in Baltimore, based upon a coincidence technique patented earlier by Rosenstock.[12] While the initial instrument used electron-impact ionization,[13] subsequent modifications included ionization by x rays[14] and vacuum UV photons. In the latter instrument (used by this author in the late 1960s), vacuum UV photons produced by a mercury lamp were selected by wavelength using a McPherson grating monochromer and transmitted into an ionization region consisting of two extraction grids with voltages of opposite polarity (Figure 4.4). At wavelengths corresponding to the photoionization threshold, each photon produced a positive molecular ion and an electron:

$$h\nu + AB \rightarrow AB^+ + e^-$$

The electron was extracted in one direction by the positive grid, detected by an electron multiplier, and passed through a low-level discriminator. The output pulse of the discriminator then initiated a linear voltage ramp in the time-to-pulse height converter. At the same time, the positive ion was extracted and accelerated in the opposite direction, where it traveled the length of the drift tube. The ion was recorded on a second detector, with the amplified pulse again passed through a low-level discriminator and used to terminate the voltage ramp. The time-to-pulse height converter then output a pulse whose amplitude was equal to the height of the voltage ramp at its termination. The multichannel analyzer then assigned this pulse to a particular channel based upon its amplitude, and the mass spectrum was obtained by accumulating the pulses from many time-of-flight cycles. Spectra could also be

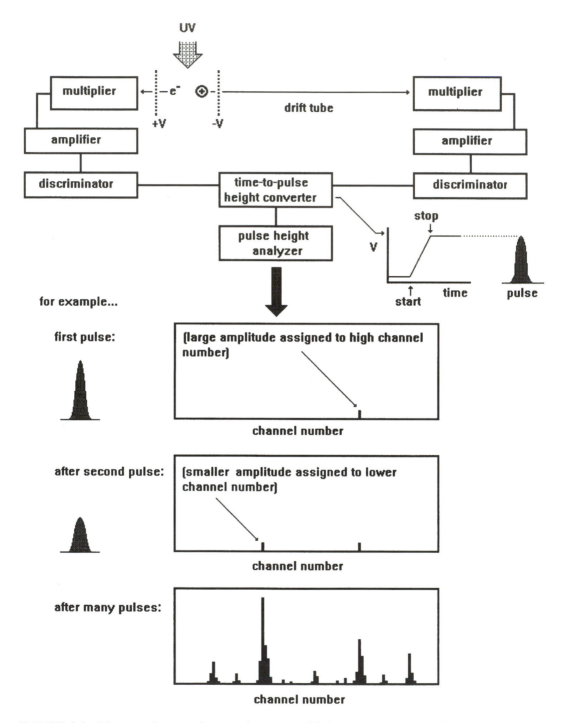

FIGURE 4.4 Schematic diagram of a coincidence time-of-flight mass spectrometer. In contrast to current TDCs, the TPHC is a single-stop device that outputs pulses, whose amplitudes are proportional to flight times and are assigned to an appropriate channel by a PHA.

obtained at other wavelengths to provide information on electronic energy states of the ion, or to study fragmentation processes:

$$hv = AB \rightarrow A^+ + B + e^-$$

Because the flight time of each ion is the time interval that begins with detection of an electron released in the same ionization event, this is a coincidence instrument. The only requirement for such instruments is that ionization events occur (predominantly) at intervals longer than the time-of-flight measurement.

The Time-to-Digital Converter (TDC).   Macfarlane's 1973 mass spectrum of sodium acetate was also recorded using a time-to-amplitude (pulse height) converter and multichannel analyzer, with one of the fission fragments providing the signal to initiate the voltage ramp. Unlike the photoionization instrument described above, however, a single ionization event may produce a multiplicity of ions (typically from 1 to 10). Because the TPHC–MCA scheme is essentially a *single-stop* approach, it discriminates against higher mass ions which arrive at the detector at a later time. For this reason, multistop time-to-digital converters (TDCs) were developed, which can generally record up to 15 stop pulses for each start pulse. By the mid 1980s time resolution for TDCs was (typically) 0.5 to 1.0 ns, providing considerably better timing accuracy than could be obtained from the 100-Msample/s (10 ns) transient recorders then available for digitizing analog signals.[4] For a time, this provided a distinct advantage for instruments utilizing coincidence techniques, particularly for instruments using high acceleration voltages and short flight tubes in which the flight times were short.

The dead time for TDCs ranges from 20 to 100 ns, which implies that some ions will be missed. Nonetheless, the TDC approach to time-of-flight measurement generally achieves the multichannel advantage for recording all ions, provided that only a few ions are produced infrequently. For example in the BIO-ION instrument, start signals occur (randomly) at approximately 1500 times/s, or once every 0.67 ms. Depending upon the mass range recorded, flight times will range between 4 to 8 μs, or about 1/100 of the average cycle time. While this would appear to result in considerable measurement dead time, the factor of 100 is necessary for random ionization events in order to keep erroneous *start–stop* correlations below 1%. In instruments in which ionization is triggered at regular intervals (such as those reported by Standing's group and discussed in Chapter 5), it is possible to achieve a much higher mass measurement duty cycle.

The Constant Fraction Discriminator (CFD).   In the low-level discriminators used in early instruments, a timing signal is generated when (and if) the incoming pulse exceeds a predetermined amplitude threshold. As shown in Figure 4.5a, an incoming pulse **3** whose amplitude is below that threshold does not generate a timing signal. However, the other two pulses (**1,2**) shown do generate timing signals, but at different times (t_1, t_2) that depend upon their amplitudes. This error in generating timing signals is significant only when the incoming pulse width (derived from the output of a multiplier or channelplate detector) is wider than the desired timing accuracy.

(a)

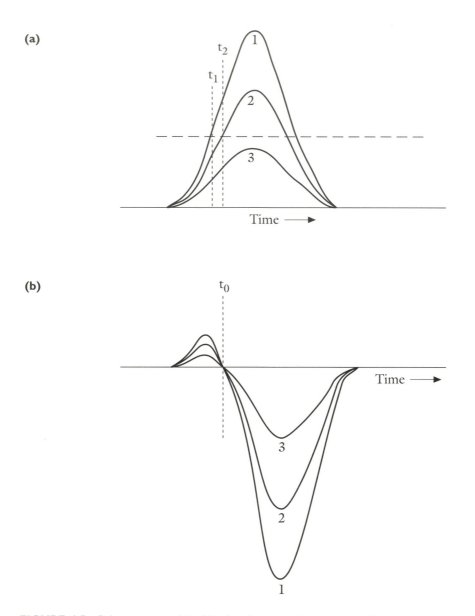

(b)

FIGURE 4.5   Pulses processed by (a) a low-level discriminator, and (b) a constant-fraction discriminator.

Because the incoming pulses generally have the same shapes (and rise times), it is possible to generate timing signals at some fraction of their full height that is independent of amplitude. As shown in Figure 4.5b, this is accomplished by summing the incoming pulse with a time-delayed inverted pulse, whose amplitude can be adjusted to determine the fraction of full height that will generate the timing signal. Thus, pulses **1,2** generate timing signals at t_0, while pulse **3** is again below the discrimination level. These constant fraction discriminators (CFDs) are used in most PDMS instruments and have timing resolutions of the order of 100–200 ps.[4]

Mechanisms of Ion Formation

In PDMS, intact molecular ions are produced by bombardment of the sample with massive, high-energy (approximately 1 MeV/nucleon) particles. Considering the fact that the 70-eV electrons used for electron impact (EI) ionization produce considerable fragmentation, and that chemical ionization (CI) requires gas-phase reactions between reagent ions and analyte molecules that are exothermic by only a few or a fraction of an electron volt, this is indeed surprising. Other desorption techniques also utilize excitation energies that exceed ionization potentials by several orders of magnitude. SIMS (secondary ion mass spectrometry) and FAB (fast atom bombardment) use primary beams with energies in the 5- to 35-keV range, while laser desorption methods use pulsed photon beams with power densities in the range of 1 to 10 MW/cm^2. The resulting spectra are generally characterized by the same even-electron, protonated molecular species found in CI spectra, suggesting that they all share some mechanistic similarities.

Ionization of a neutral molecule in the gas phase by electron impact is a relatively simple, two-body problem, in which the incoming electron kinetic energy (generally 70 eV) is accounted for by the ionization potential of the analyte molecule, the increase in internal energy of the molecule, and the kinetic energies of the products. Although only a fraction of the electron kinetic energy is deposited in the molecule, dissociation will easily occur if this exceeds its ionization potential. In contrast, the desorption of ions from the condensed (solid or liquid) state is a far more complex process that includes: (1) the initial interaction of the primary particle (or photons) with the surface or substrate, (2) the series of (energy isomerization) processes by which intact molecular species are released without appreciable fragmentation, and (3) the processes by which molecules become ionized. While numerous mechanisms have been proposed for each of these stages, that has not led as yet to a unified theory of desorption. However, agreement on the details is generally not necessary to achieve a basic understanding of the kinds of things that tend to improve desorption–ionization efficiency.

Interactions of Energetic Particles with a Surface.   The rate at which an energetic particle loses energy after entering a solid surface is known as its stopping power dE/dx, where E is the instantaneous kinetic energy of the incoming particle at a given depth x in the solid. The mechanism for transfering energy to the solid surface depends upon both the mass and the energy of the incoming particle. In this respect, it is useful to distinguish between fast, heavy ions such as the 50- to 120-MeV fission fragments used in plasma desorption and slow, light ions such as the 5- to 20-keV Cs$^+$ ions or Ar atoms used in SIMS and FAB, respectively. The interaction of fast, heavy ions with solid surfaces results in electronic excitation of the solid in the vicinity of its trajectory through the solid, known as the fission track. As shown in Figure 4.6, the electronic stopping power $(dE/dx)_e$ for ^{99}Tc ions reaches a maximum at 110 MeV (or about 1 MeV/nucleon). A second maximum is shown in Figure 4.6 for ^{99}Tc ions with kinetic energies in the keV range that results from vibrational excitation of the solid lattice, and is referred to as nuclear stopping power $(dE/dx)_N$.

From Figure 4.6 it should be obvious that energy deposition is far more efficient for the fast, heavy ions that are used in plasma desorption than for the lighter and slower ions used in SIMS techniques. More importantly, this means that more energy is deposited in the vicinity of the surface, where analyte molecules are located. This

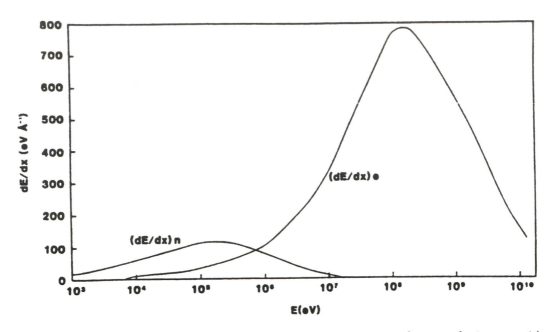

FIGURE 4.6 Nuclear $(dE/dx)_n$ and electronic $(dE/dx)_e$ stopping power as a function of primary particle kinetic energy. (Reprinted with permission from reference 1).

has several important consequences. First, it enables one to use thin foils in the PDMS technique and to place the ^{252}Cf source behind the foil. In contrast, ions in SIMS techniques generally bombard thick solid supports from the sample side. Secondly, desorption by PDMS occurs as the result of bombardment by a single, energetic fission fragment particle, while SIMS techniques use short ion bursts to achieve high-energy deposition at the surface. Not so obvious from Figure 4.6 is the fact that stopping power is increased with larger mass of the primary particle, which is the reason that xenon quickly replaced argon in the 8-keV saddle-field neutral guns that were introduced with the FAB technique. Furthermore, 35-keV Cs$^+$ ion guns have now replaced saddle-field guns altogether. Moreover, there is considerable interest in utilizing large cluster ions for ion bombardment in the keV region.

Desorption of Molecular Species.   When fast, heavy ions are used (PDMS), it is unlikely that molecular species are desorbed in the immediate vicinity of the ion track. It is rather more likely they are desorbed at some distance from the point of impact where the excitation energy density more closely matches the noncovalent attractive forces which bind the molecule to the surface, an annular region known as the interaction zone[15] (Figure 4.7). A simplistic picture of how this is accomplished assumes that the initial electronic excitation energy is *isomerized* to vibrational excitation of the solid lattice holding the sample, and that this vibrational excitation is then coupled to the weak forces holding the molecule on the surface, but poorly coupled to the internal covalent bonds of the molecule. This situation is shown in Figure 4.8 in which the substrate lattice and the internal bonds of the molecule are both represented as collections of strong oscillators, while the sample-to-surface

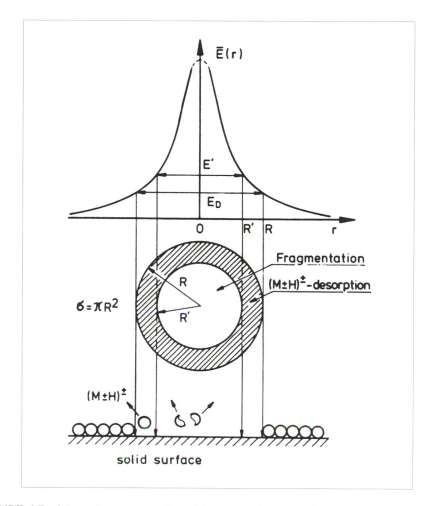

FIGURE 4.7 Schematic representation of the energy density in the region of the fission track. The shaded annular region is known as the interaction zone, from which intact molecular species are desorbed.

interactions are represented as weak oscillators. This thermal model[16] suggests that vibrational excitation of the strong oscillators in the lattice results in sufficient excitation of the binding forces to cause dissociation from the surface, and occurs in a time frame that is very short so that coupling to the internal bonds is very weak. Thus, the molecule is desorbed with very little internal energy, and is mass analyzed as a molecular species.

While a simplistic model, the thermal model does suggest that any measures that are taken to reduce the number or strength of the binding oscillators should greatly improve desorption efficiency. This is the case, and the methods which have been utilized are discussed below in the section on matrices and surfaces.

Ionization. Perhaps one of the mostly hotly debated subjects has to do with whether molecular ions observed in desorption mass spectra result from their direct

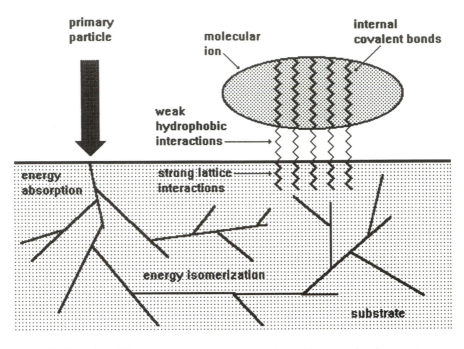

FIGURE 4.8   Thermal model for the desorption of intact molecular species.

desorption from the surface or are formed by ion-molecule reactions (basically chemical ionization) in the region just above the surface known as the *selvedge*.[17] The MK[+] species observed in IR-laser desorption have been shown[18,19] to result from gas-phase attachment reactions between K[+] ions and codesorbed neutrals. In FAB mass spectrometry, products of the extensive radiation damage of the liquid matrix may include species that lead to protonation of the analyte in the relatively high pressure region of the selvedge.[17,20] In addition, it has been shown that bispyridinium salts, which contain stable, divalent cations, are generally observed as singly charged ions that include an anionic counterion or involve electron reduction of one of the quaternary ammonium centers.[21] Such results suggest a strong tendency for charged species to be neutralized prior to desorption.

However, the opportunities for gas-phase (selvedge) reactions may be considerably reduced in PDMS in which a single projectile produces a very small number of ions in widely spaced desorption events. In addition, the number of multiply charged ions appears to increase with increasing mass, suggesting that large ions may be desorbed as easily as large neutrals. Thus, it is likely that preformed ions as well as those formed by reactions above the surface are both involved in desorption.

The Development of Matrices and Surfaces

The interaction zone is the region in which the desorption energy density most closely matches the energy required to overcome the weak forces which bind the molecule to the surface. In PDMS, a larger annular region (Figure 4.7) should be available for desorption if the surface-binding forces are reduced. That this is the case is borne out by the dramatic improvement in desorption efficiency that resulted from the introduction of the nitrocellulose surface in 1986.[22] The effect of this surface is to decrease ionic interactions to the metal surface and to promote only weak hydrophobic interactions. The resultant increase in desorption efficiency is shown in Figure 4.9a and c, which compares the spectra of porcine insulin electrosprayed onto a metal surface with that deposited on a nitrocellulose-coated surface.[23] Consistent as well with the decrease in ion interactions is the observed increase in doubly charged ion intensity, as well as the appearance of triply charged ions. A similar result can be obtained by insuring that the peptide is well folded, which reduces its contact area with the surface. At about the same time that the nitrocellulose surface was introduced, Alai et al.[24] showed that increased desorption efficiency, shifts to higher charge states, and decreased peak widths could be observed for peptides that were electrosprayed onto the sample foil in solutions of glutathione (Figure 4.9b), which has been shown to promote protein folding.[25]

A reduction in peak widths (and the consequent improvement in mass resolution) also reflects the reduction in binding energy as the ions are desorbed with considerably lower kinetic energy spreads. Thus, Silly et al.[26] have shown directly that protein folding reduces the binding energy by plotting the peak widths of two peptides (α-lactalbumin and chicken-egg-white lysozyme) as a function of pH. These peptides have nearly the same mass (14,176 and 14,307 Da, respectively) and the same tertiary structure. However, they differ in their primary structures, and hence their isoelectric points (pI = 5 and 11, respectively). In both cases, the most tightly folded forms are observed near their isoelectric points. As shown in Figure 4.10, the peak widths for their singly, doubly, and triply charged molecular ions are all minimized at their isoelectric points.

The nitrocellulose-coated foil is now commonly used in PDMS. When samples are deposited on nitrocellulose, it becomes possible to wash the sample to remove small cations (primarily Na+) which can interfere with the desorption process. While the high affinity for hydrophobic peptides is not surprising, the fact that desorption efficiency should improve with hydrophobicity and that the more hydrophobic species in a mixture of peptides could surpress desorption of less hydrophobic species was not initially obvious. Table 4.1 lists a suite of peptides, grouped by mass, whose desorption efficiencies were compared by Wang et al.[27] Figure 4.11 shows the relative desorption efficiencies of these peptides as a function of their hydrophobicities. In addition, when peptides of similar mass were analyzed in a mixture, the most hydrophobic peptide appeared to suppress the ion signals from other peptides in both the positive- and negative-ion spectra. While this may appear to create serious problems for mixture analysis, we point out that such an effect makes PDMS complementary to reversed-phase HPLC separations, which separate peptides largely on the basis of hydrophobicity. Thus, in the common situation where HPLC fractionation is used to separate peptide fragments resulting from tryptic digestion of a protein, the resulting fractions will generally contain peptides of comparable hydrophobicity, which should therefore have reasonably similar ionization efficiencies.

Additionally, the nitrocellulose surface provides an opportunity to carry out enzymatic and chemical reactions directly on the sample foil. These are discussed below, and in more detail later in Chapter 10, and include carboxy and aminopeptidase reactions that can be used to produce peptide ladder mixtures for determining amino acid sequence or the location of post-translational modifications. Following the on-foil reaction, the enzyme can be washed off prior to PDMS analysis of the resulting products.

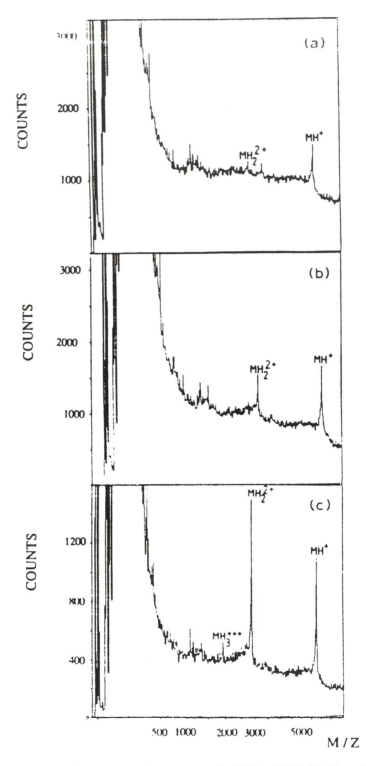

FIGURE 4.9 Positive ion PDMS spectra of porcine insulin (MW = 5778). (a) 17 nmol of electrosprayed insulin, (b) 3.5 nmol of insulin electrosprayed with 200 nmol reduced glutathione, and (c) 0.8 nmol of insulin adsorbed to nitrocellulose from a 0.1% trifluroacetic acid solution. (Reprinted with permission from reference 23).

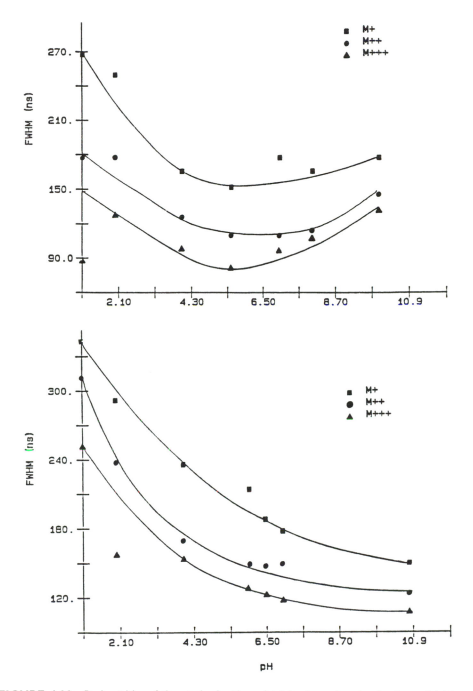

FIGURE 4.10 Peak widths of the singly, doubly, and triply charged molecular ions of (a) lactal-bumin and (b) lysozyme, as a function of pH. (Reprinted with permission from reference 26).

TABLE 4.1   Peptides grouped by numbers of amino acids.[a]

Peptide	MW	Hydrophobicity[b]
Thyrotropin releasing hormone	848.9	+443
[Arg^8]-vasopressin	1084.3	+107
[Tyr^8]-bradykinin	1076.3	–94
Dynorphin$_{1-9}$	1137.4	–262
Des-Asp^1-angiotensin I	1181.4	–544
Katacalcin	2436.7	+270
Human β-melanocyte stimulating hormone	2661.0	+160
Human parathyroid hormone$_{28-48}$	2148.5	+30
Bovine adrenal medulla docosapeptide	2839.4	–86
Human parathyroid hormone$_{39-68}$	3285.8	+217
C-peptide$_{3-33}$	3020.4	+101
Corticotropin releasing factor antagonist	3826.8	–77
Growth hormone releasing factor$_{1-29}$	3473.2	–160
Human growth hormone releasing factor$_{1-40}$	4544.3	+102
Human adrenocorticotropic hormone	4541.3	+8
Teleost fish urotensin	4869.7	–79
Human Tyr-corticotropin releasing factor	4920.9	–140

[a] Adapted with permission from reference 27.

[b] Cal/mole, calculated according to Bull, H.B.; Breese, K., *Arch. Biochem. Biophys.* 161 (1974) 665.

Peptide and Protein Analysis

PDMS provided the first real opportunity to obtain mass spectra of large biological molecules. As the first in a series of desorption–ionization techniques that produced intact molecular-ion species, its introduction initiated many of the mass-spectral strategies (based primarily on molecular weight measurements rather than fragmentation) that are now utilized in MALDI and ESI instruments. These strategies, which utilize tryptic mapping, ladder sequencing, in situ enzymatic digestion, etc., have been used to address a wide range of structural problems in the biological sciences and are described briefly below.

Interspecies Comparisons and Point Mutations.   Many of the earliest PDMS measurements involved simple interspecies comparisons, such as that shown for bovine, porcine, and human insulin in 4.12. These spectra were obtained prior to the introduction of the nitrocellulose surface, and required ion counting times from 5 to 8 hours.[28] Because such comparative measurements could produce a high degree of mass accuracy that could be utilized to reveal a single-point mutation, Alai[29] used PDMS to verify the substitution of His for Asp in the insulin of a patient suffering from familial insulinemia resulting from the substitution of *GAC* for *CAC* in the gene encoding for insulin.

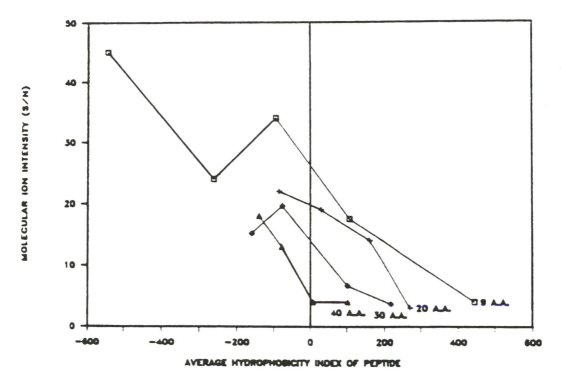

FIGURE 4.11   Relative intensities of peptides containing (a) 10, (b) 20, (c) 30, and (d) 40 amino acid residues as a function of their hydrophobicities. (Reprinted with permission from reference 27).

While simple molecular weight measurements do not provide information on the amino acid sequences of peptides, they have been profoundly useful for verifying sequences which are inferred from the nucleic acid sequences of the genes encoding the peptide. In particular, they have been used to verify the sequences of peptides produced by recombinant techniques or by total chemical synthesis, or to reveal possible post-translational modifications. More specific information, however, can be obtained by comparative mass mapping of tryptic (or other enzymatic) digests. This approach is particularly useful when the molecular-ion mass exceeds the mass range of the plasma desorption technique.

Peptide Mapping and the Location of Disulfide Bonds.   One approach to verifying that the peptide products expressed using recombinant techniques are identical to the native peptides involves comparison of their fractionated tryptic digests using reversed-phase HPLC, and is generally known as tryptic mapping. Mass spectrometry provides an opportunity for mapping enzymatic digests without fractionation. Unlike mapping techniques using HPLC, which require prior determination of the retention times of peptide fragments from native peptides, mass mapping provides an unambiguous assignment of peaks observed in these survey spectra using the expected masses that can be calculated from the amino acid sequence. For example, Figure 4.13 is the plasma desorption mass spectrum of unfractionated peptides resulting from tryptic digestion of recombinant human growth hormone (rhGH). The sequences

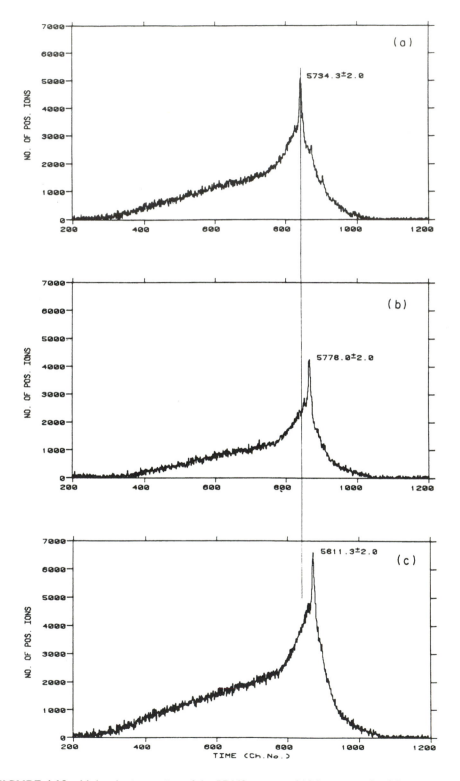

FIGURE 4.12   Molecular ion region of the PDMS spectra of (a) bovine insulin, (b) porcine insulin, and (c) human insulin. (Reprinted with permission from reference 28).

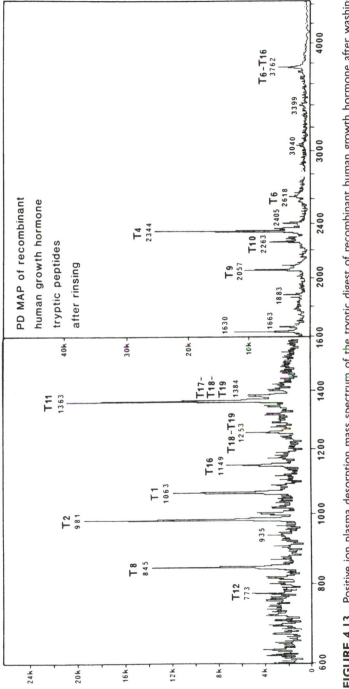

FIGURE 4.13 Positive ion plasma desorption mass spectrum of the tryptic digest of recombinant human growth hormone after washing with deionized water. (Reprinted with permission from reference 30).

TABLE 4.2   Sequences of rhGH tryptic peptides.[a]

Tryptic fragment	Position	Sequence	Calculated MH+
T1	1–9	*MFPTIPLSR*	1062.3
T2	10–17	*LFDNAMLR*	980.2
T3	18–20	*AHR*	383.4
T4	21–39	*LDQLAFDTYQEFEEAIPK*	2343.6
T5	40–42	*EQK*	404.5
T6	43–65	*YSFLQNPQTSLCFSESIPTPSNR*	2617.9
T7	66–71	*EETQQK*	762.8
T8	72–78	*SNLELLR*	845.0
T9	79–95	*ISLLLIQSWLEPVQFLR*	2056.5
T10	96–116	*SVFANSLVYGASDSNVYDLLK*	2263.5
T11	117–128	*DLEEGIQTLMGR*	1362.6
T12	129–135	*LEDGSPR*	773.8
T13	136–141	*TGQIFK*	693.8
T14	142–146	*QTYSK*	626.7
T15	147–159	*FDTNSHNDDALLK*	1490.6
T16	160–168	*NYGLLYCFR*	1149.4
T17	169–169	*K*	146.2
T18	170–173	*DMDK*	507.6
T19	174–179	*VETFLR*	764.0
T20	180–184	*IVQCR*	617.8
T21	185–192	*SVEGSCGF*	785.0

[a] Adapted with permission from reference 28.

and calculated masses of the tryptic peptides are listed in Table 4.2. The spectrum was obtained after washing the sample on the nitrocellulose foil to remove the smaller and less hydrophobic peptides. An important feature of this spectrum is the appearance of the peak labeled T6-T16, which reveals a disulfide bridge formed between Cys-54 and Cys-166.[30]

Peptide mapping has also been used to characterize abnormal variants in human hemoglobin.[31,32] Table 4.3 lists the amino acid sequences, calculated masses, and masses observed using PDMS for the tryptic peptides from the α-chain of normal human hemoglobin. In this case, peptides were purified by reversed-phase HPLC prior to mass-spectral analysis, and a human α-chain variant (Hb Bobbi) had been identified in which the HPLC peak corresponding to αT6 eluted earlier than expected. A comparison of the PDMS spectra shown in Figure 4.14 for the normal and variant αT6 revealed a mass difference of 22 Da, which corresponds to an Asp → His substitution, or a single substitution (G → C) in the codon.[30]

From Table 4.3 it is interesting to note that many of the small peptides, i.e., αT7, αT8, and αT10 were not observed by plasma desorption mass spectrometry. They were not observed as well in the HPLC map, reflecting the fact that there is generally incomplete digestion by trypsin when cleavage sites (arginine and lysine) are located in close proximity. One consequence of this incomplete digestion is the

TABLE 4.3 PDMS Analysis of AE-α^ Tryptic Peptides Purified by RP-HPLC.[a]

Peptide	Position	Sequence	Average mol. wt.	Measured mol. wt.
αT1	1–7	*VLSPADK*	728.8	729.0
αT2	8–11	*TNVK*	460.5	460.6
αT1,2	1–11	*VLSPADKTNVK*	1171.4	1171.7
αT3	12–16	*AAWGK*	531.6	531.7
αT4	17–31	*VGAHAGEYGAEALER*	1529.6	1530.2
αT5	32–40	*MFLSFPTTK*	1071.3	1071.9
αT6	41–56	*TYFPHFDLSHGSSAQVK*	1834.0	1834.2
αT7	57–60	*GHGK*	397.4	N.D.[b]
αT8	61–61	*K*	146.2	N.D.
αT9	62–90	*VADALTNAVAHVDDMPNALSALSDLHAHK*	2997.3	2999.2
αT10	91–92	*LR*	287.4	N.D.
αT10,11	91–99	*LRVDPVNFK*	1087.3	1087.7
αT11	93–99	*VDPVNFK*	817.9	818.9
αT12	100–127	*LLSHCLLVTLAAHLPAEFTPAVHASLDK*[c]	3011.6	N.D.
αT12.1	100–104	*LLSHC*[c]	614.8	615.1
αT12.2	105–127	*LLVTLAAHLPAEFTPAVHASLDK*	2414.8	2416.3
αT13	128–139	*FLASVSTVLTSK*	1252.5	1252.7
αT14	140–141	*YR*	337.4	337.4

[a] Adapted with permission from reference 30.

[b] Not determined.

[c] C as S-aminoethylated cysteine.

observation of peaks (in both the HPLC and PDMS spectrum) of peaks corresponding to αT1,2 and αT10,11. For the same reason, the peaks corresponding to T18-T19 and T17-T18-T19 are observed in the PDMS map of unfractionated tryptic peptides of rhGH shown in Figure 4.13. Thus, while these small peptides are not observed directly, it is generally possible to account for their presence. Additionally, Table 4.3 also illustrates that there are occasional cleavages at sites other than arginine and lysine. Thus, cleavage at Cys-104 results in the fragments αT12.1 and αT12.2. This occurs as well for many of the other enzymes that are commonly used to digest proteins and peptides. For example, endoproteinase Arg-C will also cleave between two adjacent lysine residues, while endoproteinase Glu-C will cleave (though incompletely) at aspartic acid residues.

Ladder Sequencing. Because plasma desorption mass spectra commonly reveal only molecular ions, it has become common to obtain amino acid sequences using carboxy or amino-peptidases to generate a mixture of truncated peptides whose mass differences correspond to the amino acid residues. One example of this is shown in Figure 4.15, in which bradykinin has been digested directly on the nitrocellulose foil with carboxypeptidase Y.[33]

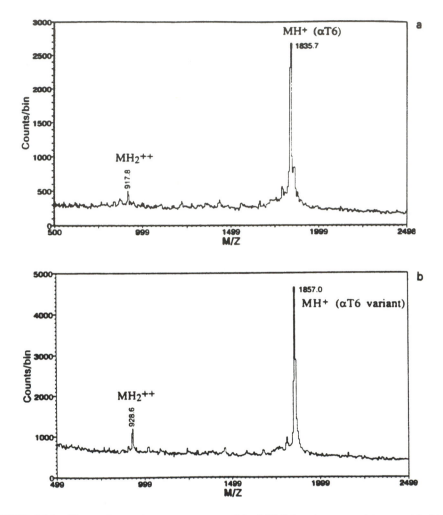

FIGURE 4.14   Plasma desorption mass spectra of the HPLC fractions containing peptide (a) αT6 and (b) an αT6 variant. (Reprinted with permission from reference 31).

Fragmentation.   Such strategies reflect the general impression that plasma desorption mass spectrometry produces only molecular weight information, but this is not always the case. It is well known that extensive fragmentation occurs in PDMS, and that for peptides this occurs primarily in the flight tube where it results in detection of fragment ions at flight times corresponding to their molecular mass. Recently, however, Bunk and Macfarlane[34] have shown that a significant amount of fragmentation occurs promptly, and can be revealed by vertically expanding portions of the mass spectrum. Their expanded spectrum of papain (MW = 23,422) shown in Figure 4.16 illustrates that sequence-specific fragmentation can be obtained for even relatively large peptides.

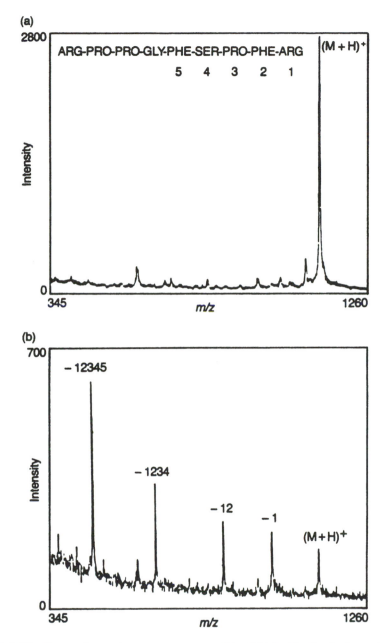

FIGURE 4.15 Plasma desorption mass spectrum of bradykinin (a) adsorbed on nitrocellulose and (b) after incubation on the foil with carboxypeptidase Y. (Reprinted with permission from reference 33).

Postscript

For many years, the major limitation of mass spectrometry was the requirement that samples be volatile. During the years that electron impact (EI) and chemical ionization (CI) were the only major methods for analyzing organic molecules, a

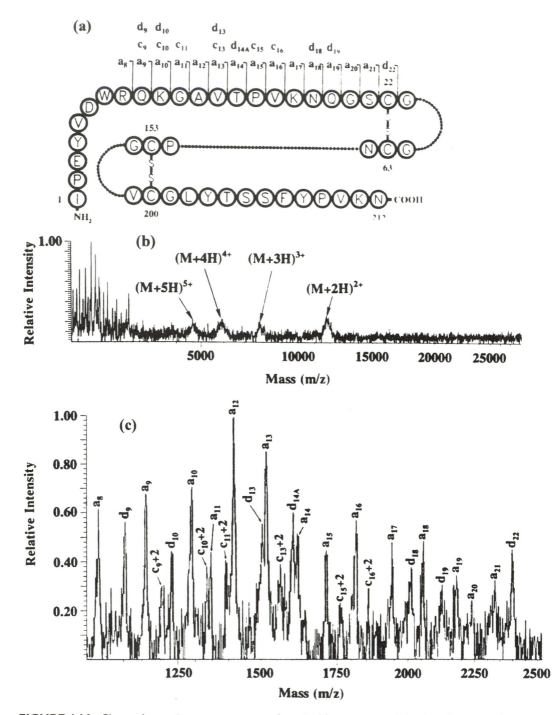

FIGURE 4.16 Plasma desorption mass spectrum of papain. (a) structure and identity of sequence fragments, (b) mass spectrum showing the multiply charged molecular ions, and (c) expansion of the spectrum in the range *m/z* 1000–2520. (Reprinted with permission from reference 34).

number of procedures were developed for converting nonvolatile compounds to their more volatile methyl or acetyl ester or silyl derivatives. Field desorption (FD) enabled the direct ionization of underivatized, nonvolatile molecules, but was a difficult technique that was never widely used by mass spectroscopists and was limited by the mass range of sector mass spectrometers.

The introduction of plasma desorption suggested that even higher mass ranges might be realized, and in a very real sense motivated much of the subsequent activity leading to the development of the new ionization techniques that are in use today. The development of static SIMS was a direct response to the PDMS method and an effort to produce the same results using light, keV primary ions from a nonradioactive source. Fast atom bombardment (FAB) utilized a liquid matrix to bring the SIMS technique to scanning sector instruments, and (ironically) considerably diminished interest in PDMS. Matrices and hydrophobic surfaces were developed for PDMS as well, and ultimately for laser desorption, giving rise to the technique now known as matrix-assisted laser desorption/ionization (MALDI), which extended mass ranges well beyond 100,000 Da. In 1990, the American Society for Mass Spectrometry recognized the contribution of PDMS in motivating the development of these techniques by granting its Award for Distinguished Contributions to Mass Spectrometry to Ronald Macfarlane.

References

1. Sundqvist, B.; Macfarlane, R.D., *Mass Spectrom. Rev.* **4** (1985) 421–460.
2. Macfarlane, R.D., *Biol. Mass Spectrom.* **22** (1993) 677–680.
3. Torgerson, D.F.; Skowronski, R.P.; Macfarlane, R.D., *Biochem. Biophys. Res. Commun.* **60** (1974) 616–621.
4. Macfarlane, R.D.; Torgerson, D.F., *Science* **191** (1976) 920.
5. Cotter, R.J., *Anal. Chem.* **60** (1988) 781A.
6. *Handbook of Chemistry and Physics* (55 Edition), Weast, R.C., Ed.; CRC Press, Cleveland, OH (1974) B9.
7. Chait, B.T.; Agosta, W.C.; Field, F.H., *Int. J. Mass Spectrom. Ion and Phys.* **39** (1981) 17.
8. Hakansson, P.; Johansson, A.; Kamensky, I.; Sundqvist, B.; Fohlman, J.; Peterson, P., *IEEE Trans. Nucl. Sci.* **NS-28-2** (1981) 1776.
9. Becker, O.; Furstenau, N.; Krueger, F.R.; Weiss, G.; Wien, K., *Nucl. Instrum. Meth.* **139** (1976) 195.
10. LeBeyec, Y.; Della-Negra, S.; Deprun, C.; Vigny, P.; Ginot, Y.M., *Rev. Phys. Appl.* **15** (1980) 1631–1637.
11. Yang, Y.M.; Sokoloski, E.A.; Fales, H.M.; Pannell, L.K., *Biomed. Environ. Mass Spectrom.* **13** (1986) 489.
12. Rosenstock, H.M., U.S. Patent 2,999,157 (1958).
13. Vestal, M.L.; Krause, M.; Wahrhaftig, A.L.; Johnston, W.H., in *Proceedings of the Eleventh Annual Conference on Mass Spectrometry and Allied Topics,* ASTM Committee E-14 (1963) pp. 358–365.
14. Krause, M.O.; Vestal, M.L.; Johnston, W.H.; Carlson, T.A., *Phys. Rev.* **133** (1964) A385.
15. Macfarlane, R.D., *Accts. Chem. Res.* **15** (1982) 15.
16. Williams, P.; Sundqvist, B., *Phys. Rev. Lett.* **58** (1987) 1031–1035.
17. Cooks, R.G.; Busch, K.L., *Int. J. Mass Spectrom. Ion and Phys.* **53** (1983) 111.
18. Van der Peyl, G.J.Q.; Isa, K.; Haverkamp, J.; Kistemaker, P.G., *Int. J. Mass Spectrom. Ion and Phys.* **47** (1983) 11.
19. Tabet, J.-C.; Cotter, R.J., *Anal. Chem.* **56** (1984) 1662–1667.

20. Fenselau, C.; Cotter, R.J., *Chem. Rev.* **87** (1987) 501.
21. Heller, D.N.; Yergey, J.; Cotter, R.J., *Anal. Chem.* **55** (1983) 1310.
22. Jonsson, G.; Hedin, A.; Hakansson, P.; Sundqvist, B.U.R.; Save, G.; Nielsen, P.F.; Roepstorff, P.; Johansson, K.E.; Kamensky, I.; Lindberg, M., *Anal. Chem.* **58** (1986) 1084.
23. Roepstorff, P.; Nielsen, P.F.; Sundqvist, B.U.R.; Hakansson, P.; Jonsson, G., *Int. J. Mass Spectrom. Ion Processes* **78** (1987) 229–236.
24. Alai, M; Demirev, P.; Fenselau, C.; Cotter, R.J., *Anal. Chem.* **58** (1986) 1303.
25. Saxena, P.; Wetlaufer, D.B., *Biochemistry* **9** (1970) 5015.
26. Silly, L.; Cotter, R.J., *J. de Physique* **C2** (1989) 37–40.
27. Wang, R.; Chen, L.; Cotter, R.J., *Anal. Chem.* **62** (1990) 1700–1705.
28. Sundqvist, B.; Kamensky, I.; Hakansson, P.; Kjellberg, J.; Salehpour, M.; Widdiyasekera, S.; Fohlman, J.; Peterson, P.A.; Roepstorff, P., *Biomed. Mass Spectrom.* **11** (1984) 242–257.
29. Alai, M., Thesis Dissertation, The Johns Hopkins University, Baltimore, MD (1988).
30. Chen, L.; Cotter, R.J.; Stults, J.T., *Anal. Biochem.* **183** (1989) 190–194.
31. Norregaard Jensen, O.; Roepstorff, P.; Rozynov, B.; Horanyi, M.; Szelenyi, J.; Hollan, S.R.; Aseeva, E.A.; Spivak, V.A., *Biol. Mass Spectrom.* **20** (1991) 579–584.
32. Norregaard Jensen, O.; Hojrup, P.; Roepstorff, P., *Anal. Biochem.* **199** (1991) 175–183.
33. Chait, B.T.; Chaudhary, T.; Field, F.H., in *Methods in Protein Sequence Analysis*, Walsh, K.A., Ed.; Humana Press: Clifton, NJ, 1987, pp 483–492.
34. Bunk, D.M.; Macfarlane, R.D., *Proc. Natl. Acad. Sci. USA* **89** (1992) 6215–6219.

5

SIMS Instruments

In 1962 (more than a decade before the introduction of plasma desorption mass spectrometry), Lehrle, Robb, and Thomas[1] described modifications to a Bendix time-of-flight mass spectrometer that would enable them to mass analyze secondary ions formed by bombardment with ions having kinetic energies in the kilovolt energy range. In their instrument, a primary-ion gun replaced the electron-impact source in a Wiley-McLaren mass spectrometer (Figure 5.1), and was capable of producing pulsed ion beams with energies from 0 to 2 keV and instantaneous ion currents of up to 5 μA. Primary-ion pulse widths were 0.25 to 3.0 μs, and the secondary ions were extracted using time-lag focusing.

The primary function of the instrument was to study ion-molecule reactions in the gas phase. However, by raising the primary gas pressure in the region between the ion-gun exit and the entrance to the ion source, energetic neutral beams could also be produced from charge-exchange reactions:

$$N_2^{+*} + N_2 \rightarrow N_2^* + N_2^+$$

(where the asterisk denotes the ion or neutral carrying high kinetic energy). By placing a positive potential on one of the two steering plates (E, in Figure 5.1), Lehrle, Robb, and Thomas were able to deflect the remaining primary ions away from the open source region, and thereby analyze the ionic products resulting from neutral-molecule reactions. In addition, they also provided for insertion of a solid target surface in the center of the ion source region for studying the secondary ions resulting from sputtering. In this case, they noted that neutral primary beams had a distinct advantage for sputtering of organic samples. Because deposition of such samples on the metal target produced an insulating surface, the use of neutral beams helped to considerably reduce the charging effects, which tended to lower sputtering efficiency under prolonged bombardment. Thus, their report presaged the fast-atom bombardment technique that would be introduced many years later.

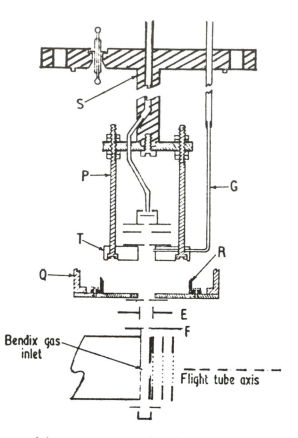

FIGURE 5.1 Diagram of the primary-ion gun described by Lehrle et al. mounted on a Bendix time-of-flight mass spectrometer. (Reprinted with permission from reference 1).

Sputtering vs. Molecular SIMS

Secondary ion mass spectrometry (SIMS) refers generally to methods in which an energetic (primary) beam of ions (or photons) is used to dislodge sample (secondary) ions from a surface for mass analysis. While the term is most often used for methods employing light primary ions with kinetic energies in the kilovolt range, most desorption techniques are in fact secondary ion mass spectrometry. Fast atom bombardment (FAB), introduced by Barber and colleagues[2] in 1981 is most certainly a SIMS technique that uses a liquid matrix as the sample surface. Although FAB initially distinguished itself as a method employing a neutral primary beam, energetic ion beams have proven equally effective in desorbing large molecules when used with the liquid matrix. Thus, the technique has often (and appropriately) been refered to as *liquid SIMS*.

Most early SIMS instruments (such as that described by Lehrle and co-workers) utilized primary-ion beams with fluxes in the 1-μA/cm^2 range or higher, which results in ablation of relatively large amounts of ions and neutrals from surface monolayers. These techniques, commonly refered to as sputtering, were used primarily for elemental analysis or for the analysis of relatively small organics. Following the introduction of plasma desorption mass spectrometry, which demonstrated the possibilities for desorbing intact molecular ions from complex biological molecules, Benninghoven and co-workers[3] introduced *static* or *molecular SIMS*. Using very short pulsed (1 ns) low flux (1 nA/cm²) beams, they were able to desorb molecular ions from nonvolatile organic compounds and amino acids that were then analyzed by time-of-flight mass spectrometry. In their nomenclature, the earlier sputtering techniques became known as *dynamic* or *atomic SIMS*.

Molecular SIMS–TOF Instruments

Plasma desorption mass spectrometry produces molecular desorption as the result of bombardment by a single, energetic heavy particle. In contrast, desorption by SIMS results from the cooperative effects of many ions in the kilovolt range delivered in a very short ion burst. Because this primary-ion pulse is considerably shorter (about 1 ns) than that used in the Lehrle instrument, pulsed ion extraction is not necessary. Thus, high voltage, constant field extraction is used, similar to that used in PDMS instruments. In addition, secondary-ion yields from each primary-ion pulse are generally low enough that time-to-digital converters can be used to record the ion flight times. Short primary-ion pulses can be produced very simply by deflecting a continuous ion beam across a narrow slit, or (more elegantly) by passing the beam through an electrostatic energy analyzer which produces a primary-ion packet of constant energy (and velocity) that does not spread in time before reaching the target.[4]

SIMS–TOF of Biological Molecules. The desorption of large biological molecules using kilovolt ion beams and time-of-flight mass spectrometry has not been a widespread approach. For the most part, the major work in this area has been carried out in Standing's laboratory at the University of Manitoba. In 1981, this group described a linear SIMS–TOF instrument (MANITOBA I) that was similar to the plasma desorption instrument of Macfarlane, except that it utilized a 2-ns pulse of Cs⁺ ions, formed by passing a continuous ion beam across a narrow slit.[5] A later instrument (MANITOBA II) incorporated a single-stage reflectron (ion mirror) and was described in Chapter 3.[6]

Among the interesting results reported by the Manitoba group was a comparison of the molecular-ion yields of a series of peptides desorbed by 8-keV Cs⁺ ions with those produced by ^{252}Cf fission fragments.[7] Their results are shown in Figure 5.2, which shows the dependence of molecular-ion yield for each method as a function of molecular mass. Because the measurements were carried out using two different instruments (at Manitoba and at Rockefeller University), the comparison was made by normalizing the molecular-ion yields of the smallest peptide (leucine enkephalin, MW = 556). The important conclusion from this study was that molecular-ion yields decrease more rapidly with increasing mass when light-ion bombardment is used.

Like plasma desorption, static SIMS generally results in mass spectra that reveal only molecular species. However, as described in the previous chapter on PDMS, structurally significant fragmentation can be observed by accumulating ion signals for long periods of time and vertically expanding the region below the molecular-ion peak. Thus, Figure 5.3 shows the mass spectrum of the peptide dynorphin (MW = 2146.2) obtained over a period of 9 hours, in which amino acid sequence ions are observed.[8]

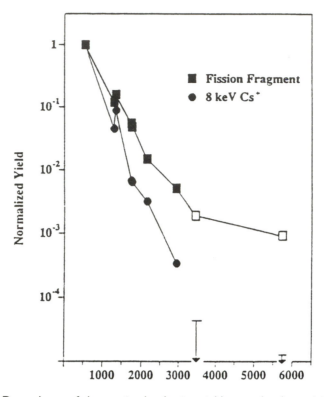

FIGURE 5.2   Dependence of the quasimolecular ion yield on molecular weight for a series of peptides bombarded with ^{252}Cf fission fragments and with 8-keV Cs$^+$ ions. (Reprinted with permission from reference 7).

Fast-Atom Bombardment, or Liquid SIMS

The technique known as fast-atom bombardment (FAB) was introduced in two simultaneous publications from the University of Manchester Institute of Science and Technology (UMIST) in 1981.[2,9] Fast atom bombardment used a neutral beam of argon atoms, formed by charge exchange reactions in a saddle-field neutral gun. Neutral beams had the important effect of reducing charging of the sample surface, but (more practically) made it possible to inject energetic particles into ion sources (such as those used on sector instruments) that were held at high positive voltage. More importantly, the FAB technique employed a liquid sample matrix (glycerol), which replenished the supply of sample molecules on the liquid surface under continuous bombardment. The ability to produce strong, continuous secondary (sample) ion currents was particularly fortuitous for scanning sector instruments, which could also provide high mass resolution. Such instruments (at the time) considerably reduced interest in methods, such as PDMS and laser desorption, that utilized lower resolution time-of-flight mass analyzers.

Colton has described FAB as a method that provides the molecular results of static SIMS under the high-flux conditions of dynamic SIMS.[10] It is now understood that the important feature of this technique is the liquid matrix, and not the neutral beam. Thus, the technique is more appropriately described as liquid *SIMS*. The liquid

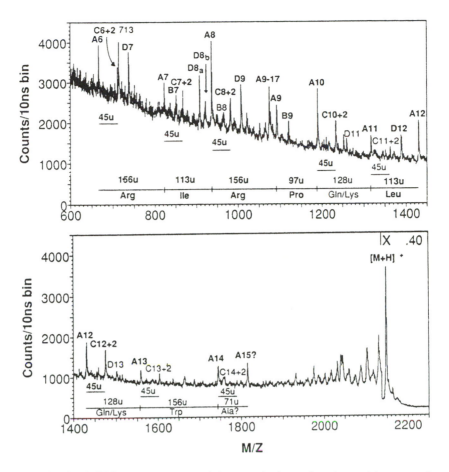

FIGURE 5.3 SIMS–TOF mass spectrum of the peptide dynorphin obtained by accumulation of the ion signal for 9 hours. (Reprinted with permission from reference 8).

matrix serves a number of important functions that include repair of the radiation damage to the surface observed in sputtering techniques, replenishment of the sample on the liquid surface, and cooling by desolvation. These have been reviewed.[11]

Liquid SIMS on a Time-of-Flight Mass Spectrometer.   Because it is a continuous technique, there might seem to be little advantage in utilizing FAB on time-of-flight mass spectrometers, which are most easily interfaced to pulsed techniques. We have noted, for example, that the first ionization method used on time-of-flight mass spectrometers was electron impact, also a continuous technique, and that there was considerable reduction in duty cycle since only a portion of the ions formed in the source were extracted by the drawout pulse. However, pulsed EI methods did provide the opportunity many years ago for interfacing gas chromatography with time-of-flight mass spectrometry, where the major problem in a very competitive GC–MS market was the absence of suitable ion recording techniques. Analogously, pulsed FAB (or liquid SIMS) offers a similar opportunity for direct interfacing of high performance liquid chromatography to the time-of-flight mass spectrometer at a time in which fast, integrated, transient recording has become available.

The possibilities for realizing online HPLC–TOF mass spectrometry motivated the development of a liquid SIMS–TOF that was reported by Olthoff et al.[12,13] and shown schematically in Figure 5.4. In this instrument, a positive pulse to the ion-gun grid (e⁻ energy) was used to produce a 1- to 10-μs, 10-μA pulse of 5-keV Xe⁺ ions focused onto a probe containing the sample in a glycerol matrix. During the ionization period, sample ions drifted into the source region where they were extracted orthogonally by a time-delayed, pulsed, multiple extraction lens system that provided space-focusing. Secondary ions were detected by a dual channelplate detector that was floated above the detector anode and provided post-acceleration of the ions to improve detection efficiency for high-mass ions. In addition, a low-mass blanking pulse was applied to a set of deflectors to minimize saturation of the channelplate detector from the intense matrix ion signal. And, because the number of secondary ions produced by a 10-μs, 10-μA pulse of primary ions was considerably larger than that produced by the 1-ns, 1-nA pulses used in static SIMS time-of-flight instruments, the ion signal was recorded using a 100-Msample/s transient digitizer.

The geometry of this instrument is particularly interesting in that it presages later approaches to interfacing continuous ionization techniques using ion storage and orthogonal extraction that are described in Chapter 7. Like the Wiley-McLaren instrument, ion currents observed at the detector increased with increasing ionization periods up to about 10 μs, where the rates of ion losses from the source begin to equal the rates of ion formation. Thus, this geometry provided a crude, passive ion storage system, in which all of the ions remaining within the ion source volume were extracted. In addition, because ion velocities were generally high in the x direction, a voltage was applied to the x-axis deflectors to focus the ions on the detector, while orthogonal extraction provided mass resolution that was not affected by the energy distribution along the x axis. Figure 5.5 shows a mass spectrum of cesium iodide ion clusters out to m/z 1200 that was obtained on this instrument. Using a continuous flow FAB[14] interface, this instrument was used successfully for online HPLC analysis of some small peptides[15] and anion-exchange chromatography (AEC) of a mixture of mono and disaccharides.[16]

The time delay utilized between the end of the 10-μs primary ion pulse and the secondary ion extraction pulse has another important consequence. It is well known that considerable fragmentation can occur in time-of-flight instruments, but is not observed in the mass spectra since it takes place in the flight tube. By delaying ion extraction from 5 to 20 μs, much of this metastable fragmentation will now occur in the source region, and will be focused at the detector at times corresponding to the fragment ion masses. Figure 5.6 shows the liquid SIMS–TOF mass spectrum of a peptide obtained after a delay time of 20 μs. Several, complete series of amino acid sequence ions are observed. Thus, such spectra more closely resemble those that would be obtained by fast-atom bombardment on a sector instrument. In addition, the spectrum was obtained by the accumulation of about 100 time-of-flight cycles, with the repetition rate about 100 cycles per second.

Imaging SIMS–TOF and Post-Ionization

While few SIMS–TOF instruments have been developed for the analysis of large biological molecules, the development of ion microprobes capable of focusing an ion

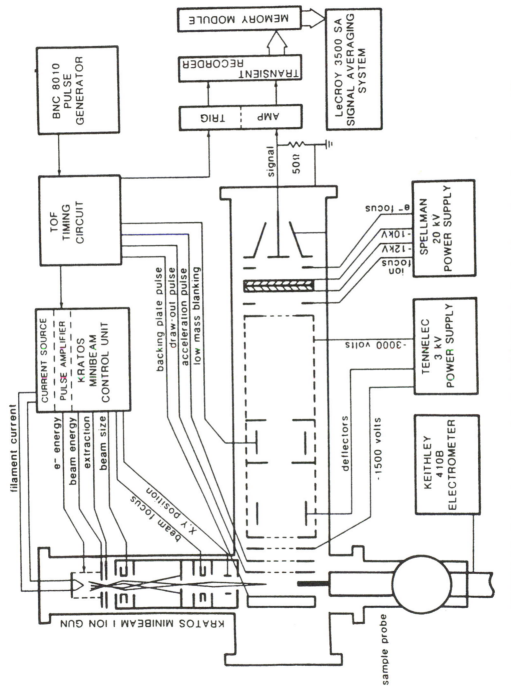

FIGURE 5.4 Schematic diagram of the liquid SIMS–TOF. (Reprinted with permission from reference 13).

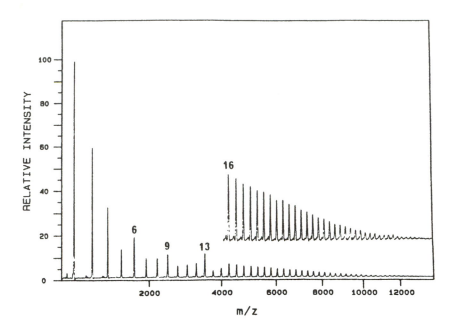

FIGURE 5.5 Mass spectrum of cesium iodide cluster ions from the liquid SIMS–TOF. (Reprinted with permission from reference 13).

YAVTGRGESPASSC

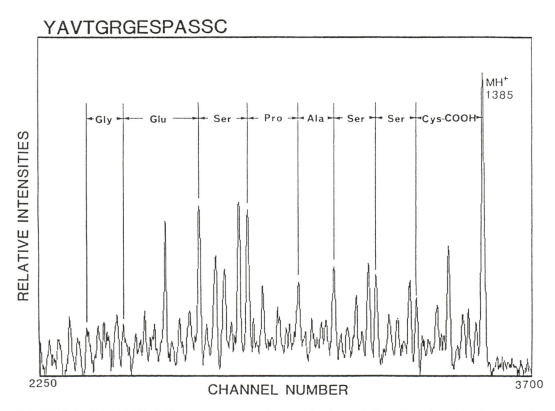

FIGURE 5.6 Liquid SIMS–TOF mass spectrum of a peptide obtained after 20 μs time delay between the ionizing and extraction pulses. The peaks marked by vertical lines belong to the y-ion series.

beam to a very small spot size[17] has made SIMS perhaps the most important method for chemical imaging of surfaces. Surface analysis by SIMS is generally accomplished by rastering the ion beam across the surface of the sample, and collecting and recording the sputtered ions in a mass spectrometer. Because it is far easier to control the position of a continuous (rather than pulsed) primary beam, quadrupole mass analyzers have been used in many surface SIMS instruments.

While early surface SIMS instruments were utilized primarily for atomic imaging, there have been a number of exciting new developments in recent years that promise to extend the capabilities for molecular imaging to biological molecules, and have increased the importance of using pulsed primary beams and time-of-flight mass analyzers. Spot sizes of less than 100 nm can now be achieved by focusing 25-keV Ga^+ ions from a liquid metal ion gun (LMIG).[18] In addition, the static SIMS techniques pioneered by Benninghoven and co-workers[3] are now being used on imaging instruments as well to provide molecular information. Because the primary-ion flux is considerably smaller, the number of secondary (sample) ions will also decrease, while the mass range to be recorded is considerably broader. Thus, the multiplex recording capability of the time-of-flight mass analyzer becomes a necessity for imaging by molecular SIMS. Figure 5.7 shows the positive-ion, secondary-ion mass spectrum of the peptide cyclosporin A (CsA; MW = 1202) dissolved in a mixture of oils used for oral drug delivery and the mass-resolved ion images of Ag^+ and $CsA+Ag^+$, and illustrates the possibilities for imaging of biological molecules.[19]

Sputtering by kilovolt ion beams produces far more neutral species than ions. When static methods are employed, one can assume that a large percentage of these sputtered neutrals will be intact molecules. Thus, there is the very real possibility for improving detection limits (sensitivity) if these neutrals can be observed by *post-ionization* in the gas phase. Such techniques are known as secondary neutral mass spectrometry (SNMS).[19]

Multiphoton ionization (MPI) using pulsed lasers is the most promising approach, and its use in imaging TOF mass spectrometers has been pioneered by Winograd and co-workers.[20] Figure 5.8 shows a generalized diagram of a SIMS–TOF imaging system with laser post-ionization. In this scheme, the post-ionizing laser beam is defocused to insure that it intersects the entire volume of desorbed neutral species. Since this presents a broad spatial distribution to the time-of-flight analyzer, dual-stage extraction optics are generally utilized to provide space-focusing. It is interesting to note that this broad ionization region results in loss of information on the spatial origin of the ions. However, spatial information is obtained by synchro-nizing the spectral recording (and display) with the rastered beam from the LMIG (ion microprobe), and not from the extraction optics (ion microscope). Also, ions desorbed directly from the surface will have higher kinetic energies than those formed within the extraction region. Thus, the reflectron voltage can be set below that of the sample stage so that directly desorbed ions pass through the back of the reflectron. One can then obtain a spectrum of the neutral species alone, and fragmentation can be controlled by the laser power density. For example, Figure 5.9 shows the mass spectra of benzo-α-pyrine (BAP) obtained from directly desorbed ions (laser off) and by post-ionization at two different laser power densities.[21]

Imaging by laser SNMS is very much an emerging technique that is not without its difficulties. Generally two or more photons must be absorbed by a molecule to enable it to reach the ionization continuum, so that the laser power density must be very high. While this is not a particular problem for atomic species, the possibilities for absorption of additional photons can result in considerable fragmentation of

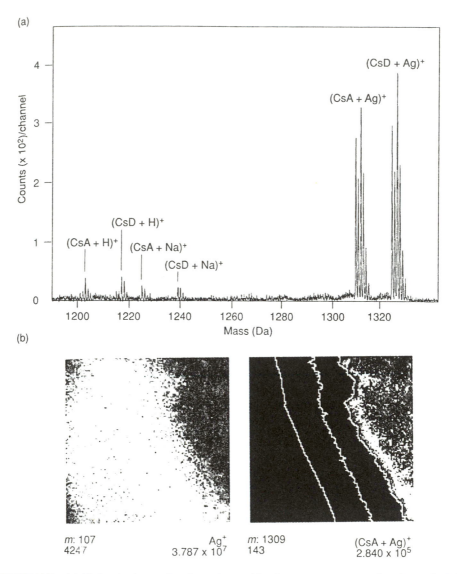

FIGURE 5.7 (a) Molecular-ion region from the positive-ion mass spectrum of cyclosporin A. (b) Imaging of Ag^+ and $CsA+Ag^+$. (Adapted with permission from reference 19).

molecules. Resonant multiphoton ionization schemes can be carried out with considerably lower power densities than nonresonant MPI, but require different (and sometimes two-color) schemes for each molecular species. Nonetheless, initial work by the Winograd group[21] is most promising, and Benninghoven et al.[22] have shown considerable reduction in fragmentation using femtosecond laser pulses.

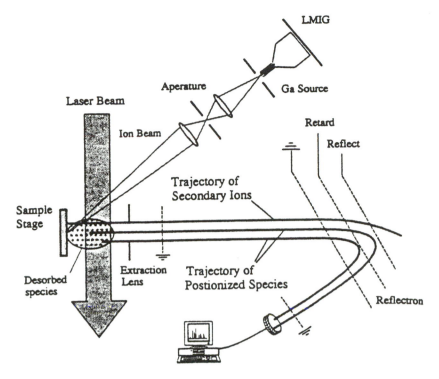

FIGURE 5.8 Schematic diagram of a SIMS–TOF imaging system with post-ionizing laser. (Reprinted with permission from reference 20).

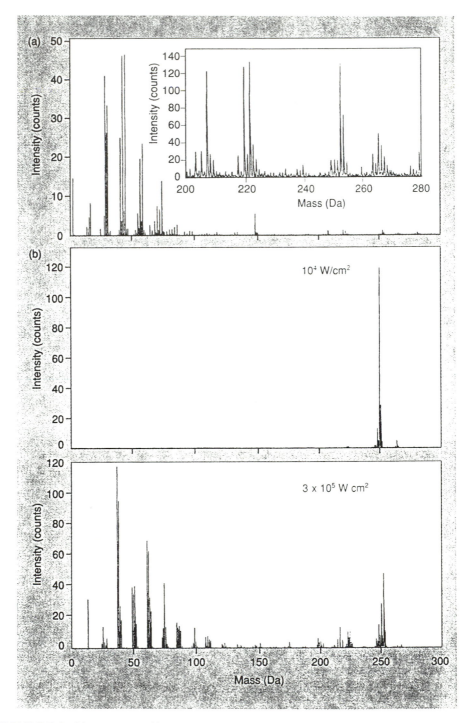

FIGURE 5.9 Mass spectra of benzo-α-pyrene: (a) SIMS spectrum with no laser post-ionization, (b) spectrum obtained by post-ionization at 280 nm and 10^4 W/cm^2, and (c) at 3×10^5 W/cm^2. (Reprinted with permission from reference 21).

References

1. Lehrle, R.S.; Robb, J.C.; Thomas, D.W., *J. Sci. Instrumen.* **39** (1962) 458–463.
2. Barber, M.; Bordoli, R.S.; Sedgwick, R.D.; Tyler, A.N., *J. Chem. Soc. Chem. Commun.* (1981) 325.
3. Benninghoven, A.; Jaspers, D.; Sichtermann, W., *Appl. Phys.* **11** (1976) 35.
4. Steffins, P.; Niehuis, E.; Friese, T.; Benninghoven, A., in *Ion Formation from Organic Solids*, Benninghoven, A., Ed; Springer-Verlag, Berlin (1983) pp. 111–123.
5. Chait, B.T.; Standing, K.G., *Int. J. Mass Spectrom. Ion and Phys.* **40** (1981) 185.
6. Tang, X.; Beavis, R.; Ens, W.; Lafortune, F.; Schueler, B.; Standing, K.G., *Int. J. Mass Spectrom. Ion Processes* **85** (1988) 43.
7. Ens, W.; Main, D.E.; Standing, K.G.; Chait, B., *Anal. Chem.* **60** (1988) 1494.
8. Poppe-Schriemer, Binding, D.R.; Ens, W.; Mayer, F.; Standing, K.G.; Tang, X.; Westmore, J.B., *Int. J. Mass Spectrom. Ion Processes* **111** (1991) 301–315.
9. Surman, D.J.; Vickerman, J.C., *J. Chem. Soc. Chem. Commun.* (1981) 324.
10. Colton, R.K., *J. Vac. Sci. Technol.* **18** (1981) 731.
11. Fenselau, C.; Cotter, R.J., *Chemical Reviews* **87** (1987) 501–512.
12. Olthoff, J.K.; Honovich, J.P.; Cotter, R.J., *Anal. Chem.* **59** (1987) 999.
13. Olthoff, J.K.; Lys, I.; Cotter, R.J., *Rapid Commun. Mass Spectrom.* **2** (1988) 171.
14. Caprioli, R.M.; Fan, T.; Cottrel, J.S., *Anal. Chem.* **58** (1986) 2949.
15. Emary, W.B.; Lys, I.; Cotter, R.J.; Simpson, R.; Hoffman, A., *Anal. Chem.* **62** (1990) 1319–1324.
16. Simpson, R.C.; Fenselau, C.C.; Hardy, M.R.; Townsend, R.R.; Lee, Y.C.; Cotter, R.J., *Anal. Chem.* **62** (1990) 248–252.
17. Castaing, R.; Slodzian, G., *J. Microsc.* (Paris) **1** (1962) 395.
18. Levi-Setti, R.; Hallegot, P.; Girod, C.; Chabala, J.M.; Li, J.; Sodonis, A.; Wolbach, W., *Surf. Sci.* **246** (1991) 94.
19. Benninghoven, A.; Hagenhoff, B.; Niehuis, E., *Anal. Chem.* **65** (1993) 630A–640A.
20. Wood, M.; Zhou, Y.; Brummel, C.L.; Winograd, N., *Anal. Chem.* **66** (1994) 2425–2432.
21. Winograd, N., *Anal. Chem.* **65** (1993) 622A–629A.
22. Mollers, R.; Terhorst, M.; Niehuis, E.; Benninghoven, A., *Surf. Interface Anal.* **18** (1992) 824.

6

Laser Desorption and MALDI

The introduction of *matrix-assisted laser desorption/ionization* (MALDI) has driven much of the current interest in time-of-flight mass spectrometry. With MALDI–TOF instruments it has been possible to determine the molecular masses of proteins weighing more than 300 kDa, while new methods based on post-source decay (Chapter 8) and tandem instruments (Chapter 9) offer the possibility for amino acid sequencing of peptides with very high sensitivity. In addition, there is the tantalizing prospect of using MALDI–TOF for rapid *ladder* sequencing of unfractionated oligomeric mixtures of oligonucleotides as an approach to sequencing the human genome. However, pulsed lasers and time-of-flight mass spectrometers have been an effective combination for a number of years. Indeed, prior to the introduction of MALDI, laser desorption had been used with some success to determine the amino acid sequences of peptides, the structures of carbohydrates and glycolipids, and the molecular weight distributions of oligomeric mixtures of industrial polymers.

By 1980, laser mass spectrometry already had a rich 20-year history, prompting an extensive review of the area by Conzemius and Capellen.[1] Reports in the 1960s included the laser vaporization of graphite[2,3] and coals,[4,5] trace elemental analysis in metals,[6] isotope ratio measurements,[7] and pyrolysis studies.[8] Generally, electron impact post-ionization was utilized for vaporization and pyrolysis studies, and (since laser pulse widths were generally of the order of a few milliseconds) pulsed extraction methods were required for time-of-flight instruments. Concurrent with improvements in laser design, the 1970s saw the development of multiphoton ionization techniques, the introduction and commercialization of laser microprobes for surface analysis, and the extension of laser mass spectrometry to organic and biological molecules using direct laser desorption of molecular ions.[9]

Laser Microprobes

In the 1970s, the availability of Q-switched Nd:YAG lasers with pulse widths in the tens of nanoseconds range enabled the design of laser time-of-flight instruments with constant extraction fields. The Nd:YAG laser has a fundamental wavelength of 1.06 μm, from which several harmonics (533 nm, 353 nm, and 256 nm) can be derived by appropriate combinations of frequency-doubling and the fundamental wavelength. The Nd:YAG laser could be focused to a very small spot size, and in 1975, Hillenkamp and co-workers[10,11] reported the design of a laser microprobe mass spectrometer. An instrument based on their design, the LAMMA 500 (LAser Micro-probe Mass Analyzer) was commercialized by Leybold-Hereaus (Köln, West Germany). While similar to imaging SIMS–TOF instruments, it was intended for the analysis of biological samples, prepared as very thin films, with the laser beam entering through the back of the sample. A later version, the LAMMA 1000, is shown in Figure 6.1. In this case, desorption is carried out from the front of the sample. While intended for the analysis of atomic (elemental) species and small organics, this instrument was ultimately used for the development of MALDI. A similar instrument, known as the LIMA (Laser Ionization Mass Analyzer), was developed by Cambridge Mass Spectrometers (Cambridge, UK).

Multiphoton Ionization

We described (in Chapter 4) an instrument designed by Vestal[12] in which volatile samples were ionized by vacuum UV photons produced by a mercury discharge source. It is particularly difficult to interface the high-pressure region of such a source to the vacuum system of a mass spectrometer, since any lens materials will absorb VUV photons. In the Vestal instrument this was accomplished by transmitting photons through a long quartz capillary. With the advent of high-powered lasers, it has become possible to carry out photoionization using two or more photons with longer wavelengths that can be transmitted through suitable lenses.

The relationship between photoionization (PI) and multiphoton ionization (MPI) schemes is shown in Figure 6.2 using benzene as an example.[13] The simplest of the single-photon processes is UV–VIS absorption spectroscopy (Figure 6.2a), which can be used to provide information on the energies of bound states of the molecule. In the example shown, the $^1B_{2u}$ state of benzene shows strong absorption at 259 nm. At much shorter wavelengths ionization will occur (Figure 6.2b). Using a particle multiplier to record total ion yield and scanning the wavelengths (which are now in the vacuum UV region), it is possible to determine the ionization potential that is observed as a sharp threshold that occurs as the internal energy of the molecule first reaches the ionization continuum. Alternatively, the wavelength can be fixed and used as the ionization source in a mass spectrometer.

The $^1E_{2g}$ state of benzene is not easily accessible by single-photon techniques. However, as shown in Figure 6.2c, it can be reached by a two-photon process using photons in the visible region of the spectrum. Such processes are possible using lasers whose focused beams provide high photon densities that enable two photons to be absorbed by the same molecule nearly simultaneously. As suggested by Figure 6.2c, information on bound states can be obtained by scanning the laser wavelength. This is usually accomplished using a suitable laser-pumped dye laser system, and a series of dyes that cover narrow regions of the spectrum. This approach is known as multiphoton spectroscopy, and has provided a wealth of information that was not previously accessible using classical spectroscopic techniques.

Multiphoton techniques can also be used for ionization. Figure 6.2d shows a simple scheme for the multiphoton ionization (MPI) of benzene using three photons,

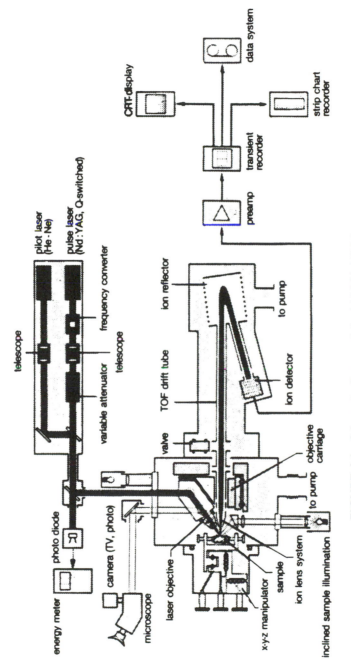

FIGURE 6.1 Schematic diagram of the LAMMA 1000 laser microprobe.

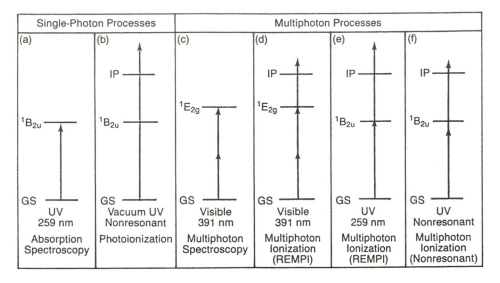

FIGURE 6.2   Examples of absorption-ionization processes for benzene. (Reprinted with permission from reference 13).

while Figure 6.2e shows an alternative scheme using only two photons. Both of these schemes proceed through a bound state, so that they are resonant techniques, that is, ionization occurs at specific wavelengths. Like single-photon techniques, MPI can be utilized for spectroscopic studies or as an ionization source for a mass spectrometer. The techniques are known as resonance ionization spectroscopy (RIS) and resonance ionization mass spectrometry (RIMS), respectively. Generally, RIS and RIMS are terms used for inorganic elemental analysis, while resonance enhanced multiphoton ionization (REMPI) is used when organic compounds are involved. In this terminology the processes shown in Figures 6.2d and 6.2e would be described as R3PI and R2PI ionization, respectively. Additionally, such schemes provide the opportunity for combining selective ionization with mass analysis. When used for elemental analysis, this can provide accurate isotope analysis in the presence of interfering isobars. Selectivity may, however, not be particularly advantageous for the mass spectral analysis of organic compounds whose identities (and, therefore, bound states) are not known a priori. In this case, multiphoton ionization can be carried out nonresonantly as shown in Figure 6.2f, but requires considerably higher power densities since it does not proceed via an intermediate bound state.

Spectroscopy and Time-of-Flight Analysis.   Multiphoton techniques are now widely used to obtain information on bound states that cannot be obtained by single-photon absorption spectroscopy. Because multiphoton processes are nonlinear, it is not practical to monitor the absorption of the laser beam intensity. Thus, a more common approach is to use laser induced fluorescence (LIF), where the absorption threshold for electronic states is obtained from the onset of fluorescence radiation. Alternatively (as we have indicated above) absorption spectra can be obtained by detecting ions produced by resonant ionization.

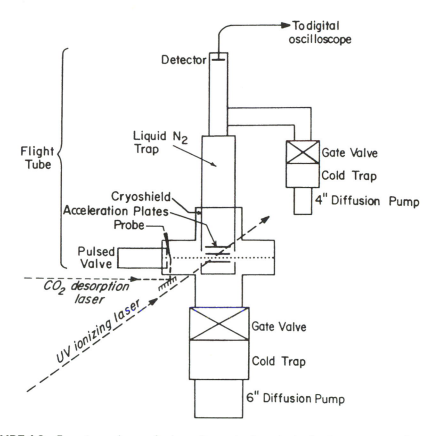

FIGURE 6.3 Experimental setup for laser desorption/laser ionization in a supersonic beam time-of-flight mass spectrometer. (Reprinted with permission from reference 17).

Li and Lubman[14] have described an interesting example in which mass analysis (using a time-of-flight mass spectrometer) in combination with vibrational–rotational cooling can be used to obtain spectroscopic information for nonvolatile compounds. In 1984, Behrson et al.[15] had recorded the origin band for indole-3-acetic acid at 293.74 nm using laser induced fluorescence spectroscopy. Later work by Levy and co-workers[16] using R2PI spectroscopy on a time-of-flight mass spectrometer placed the origin band at 285.56 nm. Because the mass spectrometer allowed them to record the ion signal corresponding to the molecular ion, they were then able to attribute the previous results as arising from decomposition products produced by thermal vaporization. Tembreull and Lubman[17] had described a supersonic beam, time-of-flight mass spectrometer in which nonvolatile compounds were desorbed by a pulsed CO_2 laser, and entrained in a supersonic jet expansion of CO_2 gas to provide rovibrational cooling. A diagram of that instrument is shown in Figure 6.3. Using that instrument they obtained the absorption spectrum of indole-3-acetic acid shown in Figure 6.4a. The shorter wavelength observed for the origin band suggested that previous measurements had been obtained for indole-3-acetic acid molecules in higher rovibronic states. Figure 6.4b shows a spectrum of 5-hydroxyindole-3-acetic acid, which is considerably more complex.

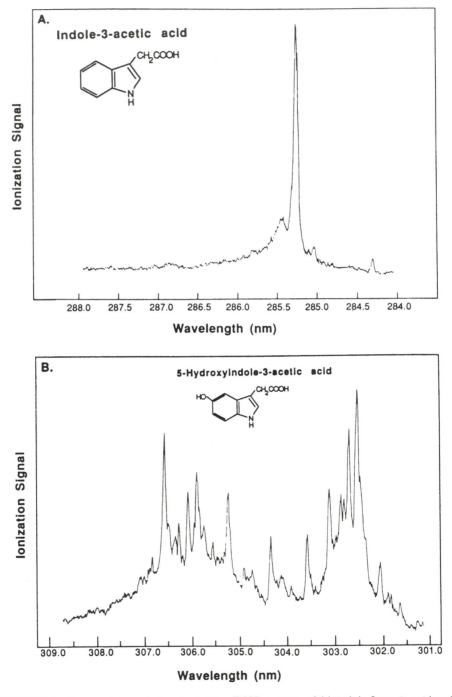

FIGURE 6.4 Resonant two-photon ionization (R2PI) spectra of (a) indole-3-acetic acid and (b) 5-hydroxyindole-3-acetic acid. (Reprinted with permission from reference 14).

In later work (which is described below), two-laser mass spectrometers using a pulsed CO_2 ablating laser, rovibronic cooling in a jet expansion, and a pulsed multiphoton ionization laser have been used successfully for obtaining mass spectra of biological molecules including peptides, small proteins, and protected oligonucleotides.

Two-Color Laser Schemes.   Multiphoton ionization can be carried out using more than one laser. In a two-laser (*two-color*) experiment, the first laser is used to saturate the excited state, while a second laser (with a different wavelength) is used to ionize the excited-state molecules. The power densities of the two lasers are often quite different. For example, Pellen et al.[18] describe the two-color multiphoton ionization of Fe atoms sputtered from a surface using a pulsed, primary-ion beam. Thus (from the last chapter), this is an example of laser secondary neutral mass spectrometry (LSNMS) which Pellen refers to as SARISA (surface analysis by resonant ionization of sputtered atoms). Using a dye laser at 302.065 nm, they resonantly excited ground-state Fe atoms using the transition $a^5D_4 \rightarrow y^5D^0_4$. Ionization from this state was then carried out using a XeCl excimer laser operating at 308 nm. Figure 6.5 shows the saturation curves for both processes. In this example, the resonant transition is saturated at considerably lower power densities than the transition between a bound and free (ionized) state. Such might not be the case, however, if the first transition required two photons. Interestingly, their instrument utilized a time-of-flight mass analyzer composed of two isochronous electric sectors, or energy analyzers and was known as an EARTOF (energy and angle refocusing time-of-flight).

Multiphoton Ionization Mass Spectrometry of Biomolecules.   While the volume of published reports concerning resonance ionization of atomic species (RIS and RIMS) and organics (REMPI), as well as nonresonant MPI, is indeed extensive, the extension of such techniques to larger, biological molecules has been much less common. The most successful approaches have been carried out in instruments (such as that shown in Figure 6.3) in which nonvolatile species are ablated by a pulsed CO_2 laser, and are entrained and cooled in a rapid jet expansion of an inert gas. The cooled neutral species are then ionized by a short-pulsed, high-powered, and tightly focused second laser to reduce the time and spatial distributions as required by the time-of-flight mass analyzer.

Peptides will generally contain one or more chromophoric residues (tryptophan, phenylalanine, or tyrosine). If the vaporized peptide molecules are not completely cooled they will have a broad absorption band in the UV region, so that R2PI ionization can be carried out at a convenient wavelength. Figure 6.6 shows the R2PI mass spectra of leucine enkephalin obtained by Li and Lubman[19] using the 266-nm output of a frequency-quadrupled, Q-switched Nd:YAG laser at two different power densities. At 1.5×10^7 W/cm² there is extensive fragmentation that can be utilized for obtaining the amino acid sequence. Fragmentation (as well as the total ionization efficiency) is reduced at 10^6 W/cm². The nucleotide bases thymine, adenosine, cytosine, and guanosine, also have strong UV absorption. Thus, Lindner and Grotemeyer[20] were able to obtain MPI spectra of a series of protected dideoxyribonucleotides using the 242.5-nm output from a tunable dye laser pumped by a frequency-quadrupled Nd:YAG laser. Figure 6.7 shows a spectrum of the protected dinucleotide d(ApTp) and an interpretation of the fragmentation.

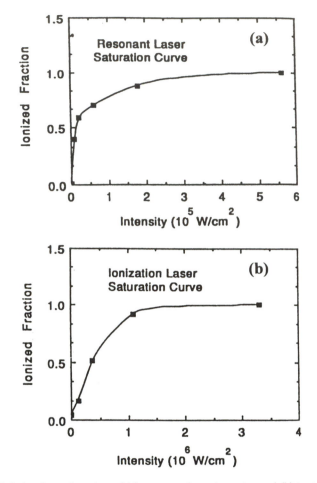

FIGURE 6.5 Fe^+ signal as a function of (a) resonant laser intensity and (b) ionizing laser intensity for Fe atoms sputtered from a Si matrix. (Reprinted with permission from reference 17).

Extension of multiphoton ionization techniques to much larger molecules has indeed been rare. The most notable example was a mass spectrum of insulin obtained by Grotemeyer and Schlag in 1987.[21] It is shown in Figure 6.8.

Laser Desorption

In 1970, Vastola, Mumma, and Pirone obtained a mass spectrum of sodium hexylsulfonate ($Na^+C_6H_{13}SO_3^-$) using a pulsed laser.[22] Their spectra recorded cation-ized salts and dimers $Na_2^+X^-$ and $Na_3^+X_2^-$ (where X = hexylsulfonate), but none of the decomposition products normally observed when these compounds were ther-mally vaporized for analysis by electron impact. This was a significant achievement since organic salts were deemed to be the compounds most intractable to mass spectral analysis. Because they observed molecular organic ions formed directly by the laser, their report is perhaps the first example of laser desorption.

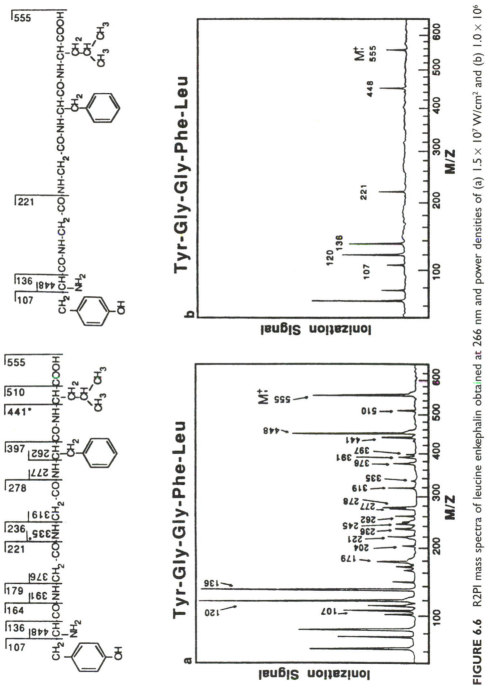

FIGURE 6.6 R2PI mass spectra of leucine enkephalin obtained at 266 nm and power densities of (a) 1.5×10^7 W/cm² and (b) 1.0×10^6 W/cm². Structures and cleavage sites are also shown. (Reprinted with permission from reference 19).

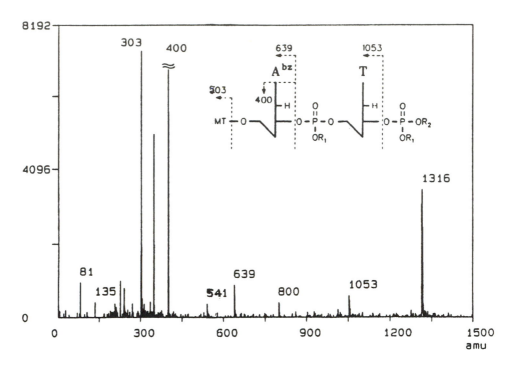

FIGURE 6.7 MPI mass spectrum of d(ApTp) (MW = 1314 Da) and assigment of the fragment ions, obtained at 242.5 nm and 5×10^6 W/cm². (Reprinted with permission from reference 20).

FIGURE 6.8 High resolution MPI mass spectrum of insulin. (Reprinted with permission from reference 21).

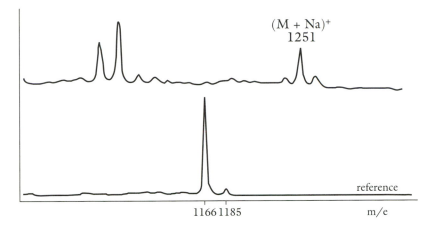

FIGURE 6.9 Infrared laser desorption mass spectrum of the molecular-ion region of digitonin and the reference compound, phosphazine. (Reprinted with permission from reference 9).

If this report went unnoticed at the time, a subsequent report by Kistemaker and co-workers[9] in 1978 most certainly caught the attention of mass spectroscopists. Using a CO_2 laser (10.6 μm) or the fundamental wavelength of a Nd:YAG laser (1060 nm), they were able to obtain molecular ions for a number of nonvolatile, high molecular weight compounds. For example, Figure 6.9 shows the molecular-ion region of the laser desorption mass spectrum of digitonin that was obtained on a magnetic-sector mass spectrometer equipped with a zoom lens. At the time, the possibility that laser desorption could be extended to other, larger biological molecules, and (in addition) might provide an alternative to the relatively difficult field desorption technique, was met with a great deal of enthusiasm. In addition, both of the lasers used by the Kistemaker group utilized radiation in the infrared region, and their results were published at a time in which there was considerable interest in the analysis of thermally intractable compounds using rapid-heating techniques. Many of these techniques utilized the activated emitters commonly used in field desorption instruments. They are reviewed here briefly.

Rapid Vaporization Techniques. Field desorption (FD) was an extension of the field ionization (FI) technique described by Inghram and Gomer[23] in 1955. In that technique, gas-phase molecules were ionized by extraction of an electron in the presence of a very high electrical field (10^8V/cm) by a process that has been described as quantum mechanical tunneling. Fields of this magnitude could be achieved by placing a thin wire (at high voltage) next to a grounded counter-electrode and activating the wire by inducing the growth of dendrites with very small radii. Because the electron tunneling mechanism did not impart significant internal energy to the ions, fragmentation was considerably less than that observed by electron impact. Field ionization, however, did require that the sample be volatile and produced (primarily) radical molecular ions: M^+. In the field desorption technique, introduced by Beckey,[24] solutions of nonvolatile samples were deposited directly on the activated emitter. In addition, the emitter was heated by passing an electrical current through the wire, and the molecular ions observed were primarily even-electron species: MH[+] and MNa[+]. The technique was difficult because of the rather lengthy process for activating emitters, and the frequency with which they would break under the stress of the high electrical field.

In the late 1970s, a number of new techniques were introduced that reflected the growing belief that the problems associated with sample volatility could be addressed by rapid thermal vaporization, rather than by extraction in high electrical fields. These led to our later understanding of the effects of pulsed laser desorption. These methods included direct EI by McLafferty and Baldwin,[25] the in-beam EI technique of Ohashi et al.,[26] flash volatilization by Daves and co-workers,[27] the activated emitter techniques of Hunt et al.[28] and Holland and co-workers,[29] and the direct CI methods of Hansen and Munson[30] and Cotter and Fenselau.[31] The formation of even-electron species by field desorption had always been inconsistent with those mechanisms invoked for field ionization, and Holland et al.[29] were able to show that desorption rates were (in fact) dependent upon heating, and independent of the field. A number of methods[26,30,31] utilized extended probes that achieved rapid heating by placing the sample directly inside the heated ion source, an approach for the analysis of nonvolatile samples that was novel at the time, but had been suggested earlier by Reed.[32] The success of these rapid heating approaches was explained by Friedman and co-workers.[33] Using the Arrhenius plots for the competing processes of vaporization and decomposition shown in Figure 6.10, they suggested that rapid heating placed nonvolatile compounds within a temperature range at which the rate of vaporization exceeded that for decomposition. The implications for pulsed lasers (in particular infrared lasers) was obvious, since these should lead to considerably higher heating rates than could be achieved by resistive heating.

Thermal Desorption.   Using rapid heating techniques, intact neutral species were vaporized. They were then ionized in the gas phase by electron impact or chemical ionization. At the time, it remained to be shown that organic ions could be produced directly by heating, that is without assistance from electron ionization or high electrical fields. Using a quadrupole mass spectrometer configured for the quantitative analysis of urinary calcium by thermal ionization, Cotter and Yergey were able to record mass spectra (Figure 6.11) of a series of quaternary ammonium salts that were dominated by the quaternary ammonium cation (R_4N^+), as well as M+Na[+] ions for some simple sugars vaporized in the presence of NaCl.[34,35] At the same time, Stoll and Rollgen showed that vaporization of intact quaternary ammonium ions from field desorption emitters occured at lower temperatures as the size of the alkyl groups (R) was increased (presumeably because the ionic bond strength between the organic cation and the Cl[-] ion was reduced),[36] and that M+Li[+] ions (in the presence of lithium salts) could be desorbed using low electrical fields that were clearly insufficient for ionization according to the field-desorption process.[37] The experiments of these two groups had important implications for the series of laser desorption reports that were to follow from a number of laboratories. Because most of these early experiments utilized infrared lasers, thermal mechanisms were generally proposed in which desorbed ions included those that were "preformed" or resulted from gas-phase reactions between codesorbed neutral species and alkali ions.

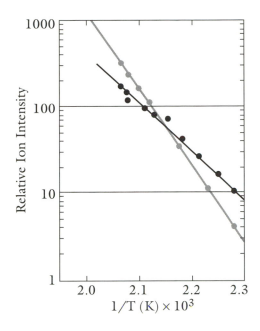

FIGURE 6.10   Relative intensity of *m/z* 363 and 235 ions from thyrotropin releasing hormone (TRH) as a function of 1/T, obtained by evaporation of TRH from a copper probe surface; (●) protonated parent molecule of PCA-His-Pro-NH$_2$, *m/z* 363, and (●) *m/z* 235 ion formed by loss of pyrrolidinoine carboxyl amide from PCA-His-Pro-NH$_2$. (Reprinted with permission from reference 33).

Infrared Laser Desorption.   Following the initial report by Kistemaker et al.,[9] Heresch et al.[38] and Cotter[39] combined pulsed IR laser desorption with sector instruments, while Vestal[40] and Stoll and Rollgen[41] utilized quadrupole mass spectrometers to carry out laser desorption experiments. Further work by Kistemaker and co-workers[42,43] established a thermal mechanism for laser desorption in which cationized species were produced by gas-phase attachment of alakali ions to codesorbed neutral species. Cationized species were also observed at much shorter wavelengths (256 nm) by Heinen,[44] while M+Ag$^+$ ions were reported by Cooks et al.[45] for solid samples deposited onto silver foils with ammonium chloride. The first laser desorption time-of-flight results were reported by Hercules and co-workers[46] using a laser microprobe instrument to observe alkali- and halide-ion attachment in positive- and negative-ion mass spectra, respectively, and by VanBreemen et al.[47]

A thermal interpretation of the mechanisms for desorption made a great deal of sense for these early laser desorption experiments. Infrared lasers are not easily focused to a small spot size, so that the 1 MW/cm^2 power densities required for the onset of desorption were achieved with instantaneous pulse energies of the order of 100 mJ (compared with the 0.1- to 1-mJ lasers used in MALDI instruments). Reports by Kistemaker et al.[42,43] and Cotter and co-workers[47,48] established that the resultant heating of the sample substrate at high laser energies (and the relatively slow dissipation of temperature) produced desorption of neutral and alkali species well beyond the actual length of the laser pulse. Using a time-of-flight instrument in which ions

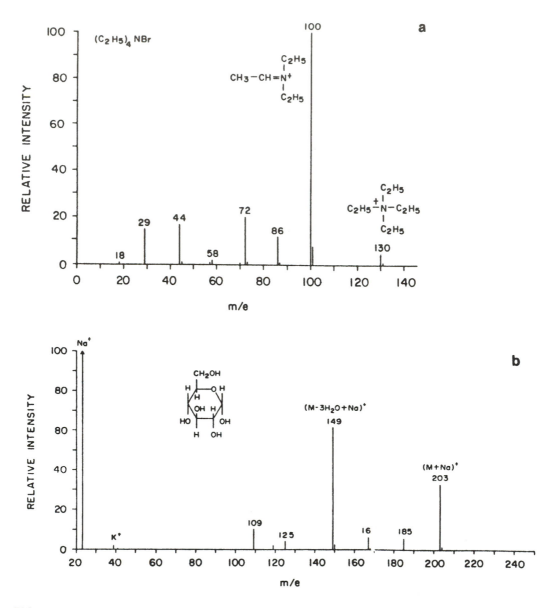

FIGURE 6.11   Thermal desorption mass spectra of (a) tetra-*n*-butyl ammonium chloride and (b) glucose with NaCl. (Reprinted with permission from references 34 and 35, respectively).

could be extracted from the source at different times after the laser pulse, Cotter and Tabet[49] were able to show that ions were desorbed with decreasing internal and kinetic energies. An example is shown in Figure 6.12, in which mass spectra were obtained for the peptide leucine enkephalin at different delay times between the laser pulse and ion extraction. As the delay time is increased, ions with decreasing internal and kinetic energy are observed as shown by the decrease in amino acid sequence-specific fragmentation and improvement in mass resolution. Infrared laser desorption

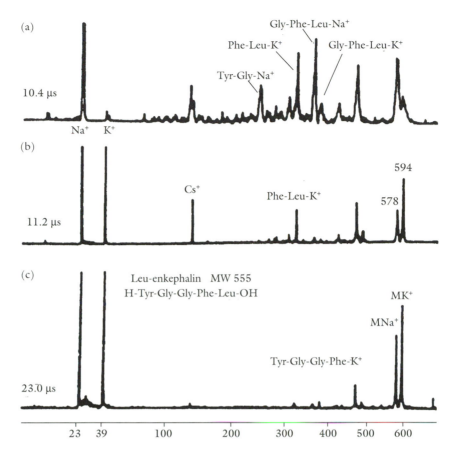

FIGURE 6.12 Laser desorption mass spectra of leucine enkephalin obtained by extracting ions from the source at 10.4, 11.2, and 23.0 μs after the laser pulse. (Reprinted with permission from reference 49).

(IRLD) in which the primary mechanism is alkali ion attachment has been particularly useful for the analysis of neutral polymers. Figures 6.13 and 6.14 show the IRLD mass spectra of an oligosaccharide[50] and an oligomeric mixture of polypropylene glycol,[51] respectively. In the former, delayed extraction of the ions produces the opportunity for considerable in-source fragmentation that can be used for structural analysis.

The thermal mechanism invoked for laser desorption mass spectrometry at the time did not suggest a particular dependence upon wavelength, in contrast to the multiphoton techniques described above. However, Hillenkamp and co-workers[52] had begun an interesting investigation on the use of UV-absorbing matrices to enhance the laser desorption of peptides. In their initial work, peptides desorbed in the presence of the amino acid tryptophan showed a small (but not particularly significant) reduction in the laser power required for the onset of desorption. However, this approach was to prove to be the right one, once the correct matrix was found.

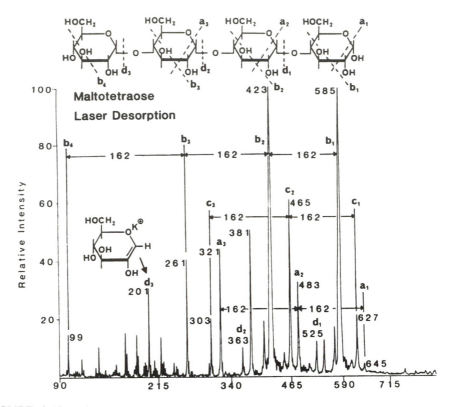

FIGURE 6.13 Infrared laser desorption (IRLD) mass spectrum of maltotetraose in KCl. (Reprinted with permission from reference 50).

Matrix-Assisted Laser Desorption

The technique, which we now know as matrix-assisted laser desorption/ionization (MALDI), was developed simultaneously in two laboratories in 1987. The first report of high mass ions (above m/z 10,000) was a paper presented by Koichi Tanaka of the Shimadzu Corporation (Kyoto, Japan) at the Second Japan-China Joint Symposium on Mass Spectrometry, held September 15–18, 1987, in Takarazuka, Japan.[53] Using a pulsed N_2 laser (337 nm) and a time-of-flight mass spectrometer equipped with a coaxial reflectron, they recorded molecular ions at m/z 34,529 from carboxypeptidase-A dissolved in a slurry of glycerol and an ultrafine metal powder. In addition, they reported a mass spectrum of lysozyme (MW = 14,307) containing multimeric ions up to the pentamer recorded at m/z 71,736 (Figure 6.15). At the same time, Michael Karas and Franz Hillenkamp from the University of Muenster (Germany) had developed a matrix-assisted technique using a frequency-quadrupled (266 nm) Q-switched Nd:YAG laser to desorb intact molecular ions from proteins dissolved in a matrix solution containing nicotinic acid.[54] Their first high mass results were reported at the International Mass Spectrometry Conference (IMSC) in Bordeaux, France in August 1988, and included molecular ions for bovine serum albumin observed in their mass spectrum at m/z 66,750 (Figure 6.16). Results from both of these groups were first published in 1988,[55,56] followed by a number of other reports by Hillenkamp and Karas for proteins with molecular weights in excess of 100 kDa.[57,58]

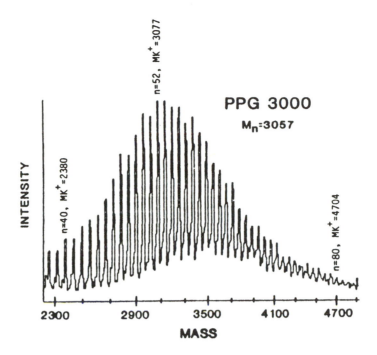

FIGURE 6.14 Infrared laser desorption (IRLD) mass spectrum of an oligomeric mixture of polypropylene glycol (PPG 3000) in KCl. (Reprinted with permission from reference 51).

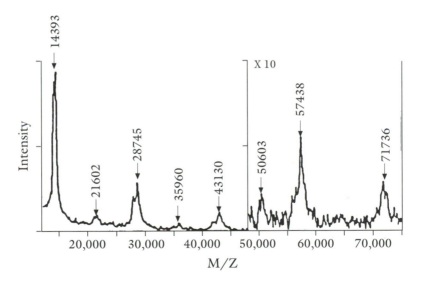

FIGURE 6.15 Laser desorption mass spectrum of lysozyme using a slurry of glycerol and ultrafine metal powder as the matrix. (Reprinted with permission from reference 53).

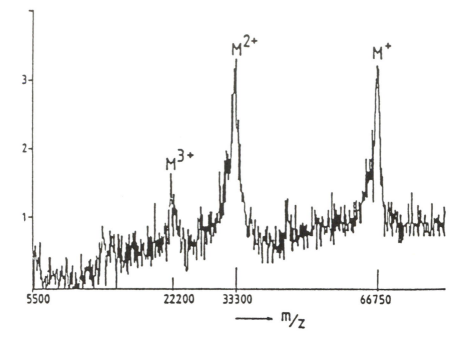

FIGURE 6.16   Matrix-UVLD mass spectrum of bovine albumin. (Reprinted with permission of reference 56).

The results reported by Karas and Hillenkamp were obtained on a commercial LAMMA 1000 (Leybold-Hereaus, Koln, Germany) laser microprobe equipped with a reflectron. In this instrument, the acceleration energy is 3 keV (considerably lower than most MALDI instruments in use today) so that a post-acceleration potential of 9 kV was added to increase the ion kinetic energy at the detector and enable recording of heavier ions.[56] The approach of Hillenkamp and Karas, using organic UV-absorbing matrices, has been the more widely adopted approach, although a number of laboratories rapidly produced similar results using a variety of other instrumental configurations, lasers, wavelengths, and matrices. In 1989, Beavis and Chait reported MALDI spectra of proteins using a linear time-of-flight mass spectrometer with dual-stage ion extraction,[59] while the group at Upsalla University in Sweden used a commercial BIO-ION Nordic (Upsalla, Sweden) plasma desorption mass spectrometer modified by the addition of an excimer-pumped dye-laser system with the output frequency-doubled to the 260–300-nm range.[60] In 1990, Spengler and Cotter obtained ions at m/z 332,000 corresponding to the dimer of β-galactosidase on a modified CVC Products (Rochester, NY) time-of-flight mass spectrometer.[61] This instrument, based upon the 1955 design by Wiley and McLaren,[62] utilized pulsed (time-lag) ion extraction and an acceleration energy of 2700 eV and had been used previously for infrared laser desorption.[47,48] In order to record ions of this size, the length of the extraction pulse was increased from 8 to 30 μs and the ions were post-accelerated to 20 keV prior to their striking the detector.

While the initial work by Hillenkamp and Karas utilized the 266-nm line (fourth harmonic) from a Nd:YAG laser and nicotinic acid as the UV-absorbing matrix, it quickly became apparent that longer wavelengths and other matrices were also effective for the MALDI technique. Beavis and Chait used the third harmonic (355 nm) from a Nd:YAG laser in combination with cinnamic acid derivatives (ferulic, caffeic, and sinapinic acid) as matrices.[63] Williams and co-workers[64] used the 581-nm wavelength from an excimer-pumped dye laser for desorption of peptides and nucleic acids from frozen aqueous solutions. Pulsed IR laser radiation at 2.94 μm (from a mechanically Q-switched Er:YAG laser)[65] and 10.6 μm (from a TEA-CO$_2$ laser)[66] have also been used successfully by Overberg and co-workers, in combination with both UV-absorbing matrices and carboxylic acids, glycerol, and urea. 2,5 Dihydroxybenzoic acid (DHB or gentisic acid) and α-cyano-4-hydroxycinnamic acid[67] are also widely used matrices, while picolinic acid[68] and combinations of picolinic acid and 3-hydroxy-picolinic acid (HPA)[69] have been introduced as effective matrices for oligonucleotides. Other, and perhaps more novel, matrices include rhodamine dyes,[70] C$_{60}$ (fullerenes),[71] and the liquid matrix, 3-nitrobenzyl alcohol (3-NBA).[72] A number of reviews and compilations of matrices have been published.[73] The chemical structures of the most commonly used matrices are shown in Table 6.1.

Chevrier and Cotter used a Photon Technology International (Ontario, Canada) 1.2-mJ, 600-ps (337 nm) pulsed nitrogen laser and caffeic acid for the MALDI analysis of peptides and proteins.[74] Pulsed nitrogen lasers are considerably simpler in design and easier to operate than Nd:YAG lasers, and are not subject to the shot-to-shot or long-term variations in pulse power that arise from effects of ambient temperature on the nonlinear crystals used for frequency doubling or tripling. They are also less expensive than Nd:YAG lasers, and for this reason are utilized in many of the commercial MALDI instruments. The 600-ps laser pulse width is considerably shorter than the 10-ns pulse width of the Nd:YAG laser described in the initial report by Hillenkamp and Karas,[56] which may provide a resolution advantage. However, a more compact 100-μJ, 3-ns pulsed nitrogen laser from Laser Sciences (Cambridge, MA) has been more common for commercial benchtop instruments.

The MALDI Technique and Its Mechanisms. In the MALDI technique, peptides or proteins are generally solublized in a 0.1% aqueous trifluroacetic acid (TFA) solution to a concentration of approximately 10^{-5} M. One microliter of this solution is mixed (usually directly on the probe tip) with a saturated aqueous solution (around 10^{-3} M) of matrix and the solvent allowed to evaporate to form crystals. The high excess of matrix to sample (100:1 to 10,000:1) is important, since the matrix serves as the primary (and highly efficient) absorber of the UV laser radiation and breaks down rapidly, expanding into the gas phase and carrying along undamaged analyte molecules. Additionally, the high matrix/sample ratio reduces associations between analyte molecules, and provides protonated and free-radical products that ionize the molecules of interest.[75] In practice, it is important to establish the threshold irradiance, the laser pulse power that results in the onset of desorption of the matrix ions. Molecular ions are generally observed at slightly higher irradiances, while higher laser power generally results in extensive fragmentation to small and structurally nonspecific ions, high ion kinetic energies, and loss of mass resolution.

A number of detailed models for the mechanisms for MALDI ionization have been described. The *homogeneous bottleneck* mechanism[76,77] attempts to explain the vibrational mismatch between the matrix and analyte that leads to relatively *cool* (low internal energy) analyte molecules, while the *cool plume* or hydrodynamic model[78] focuses on the actual expansion into the gas phase. Experimentally, Cotter and Spengler[61] determined the initial kinetic energy distributions of molecular and fragment

TABLE 6.1   Matrices used in MALDI mass spectrometry.

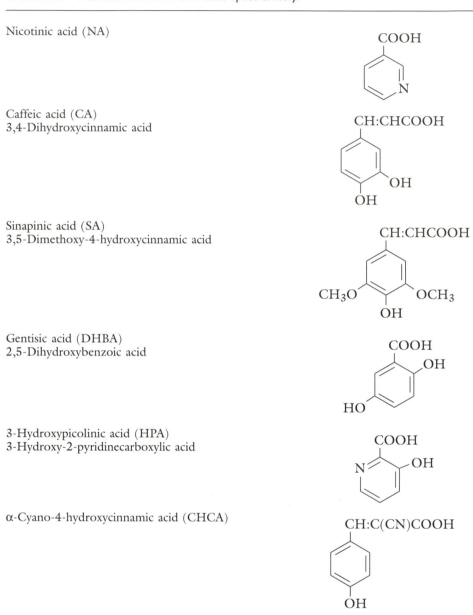

Nicotinic acid (NA)

Caffeic acid (CA)
3,4-Dihydroxycinnamic acid

Sinapinic acid (SA)
3,5-Dimethoxy-4-hydroxycinnamic acid

Gentisic acid (DHBA)
2,5-Dihydroxybenzoic acid

3-Hydroxypicolinic acid (HPA)
3-Hydroxy-2-pyridinecarboxylic acid

α-Cyano-4-hydroxycinnamic acid (CHCA)

ions from the matrix tryptophan to be of the order of 1 eV. Subsequently, three separate reports by Beavis and Chait,[79] Becker and Huth-Fehre,[80] and Pan and Cotter[81] determined that the average forward velocities and velocity distributions of analyte ions were the same as those of the matrix ions, suggesting that analyte ions are entrained in the expanding matrix plume and that the kinetic energies of analyte ions increase with increasing mass. Because the latter results in somewhat shorter flight times than would be expected for ions that are accelerated to the same final kinetic energy, this poses a significant problem for mass calibration, which is normally

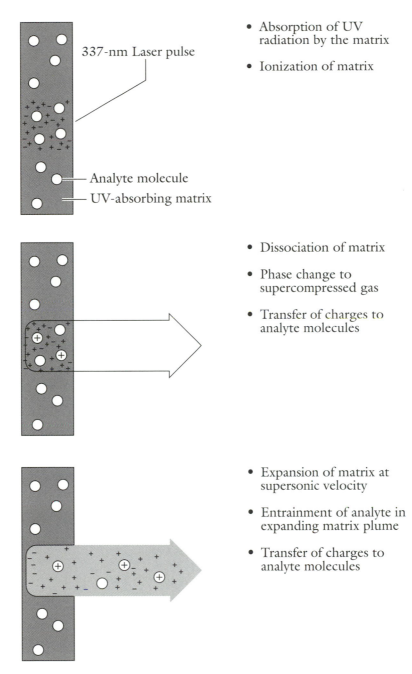

337-nm Laser pulse

Analyte molecule

UV-absorbing matrix

- Absorption of UV radiation by the matrix
- Ionization of matrix

- Dissociation of matrix
- Phase change to supercompressed gas
- Transfer of charges to analyte molecules

- Expansion of matrix at supersonic velocity
- Entrainment of analyte in expanding matrix plume
- Transfer of charges to analyte molecules

FIGURE 6.17 Schematic of the MALDI desorption process.

addressed by using calibration compounds whose masses bracket that of the ion of interest. At the same time, these results provide an approximate understanding of the events that occur during the MALDI process.

These events are illustrated schematically in Figure 6.17. In the initial stage, UV radiation is absorbed by the matrix, resulting in extensive ionization. Isomerization of this ionizing energy then results in extensive breakdown of the crystalline structure of the matrix, and dissociation of matrix molecules into small, neutral fragments and free-radical ions. Effectively, this amounts to a phase change to a supercompressed gas, which rapidly expands at supersonic velocities, entraining intact analyte ions. Charge- or proton-transfer reactions can occur at any time during this process to produce the protonated molecular ion species that are observed in the mass spectra. In addition, little internal energy is transfered to the analyte molecules, which may (in fact) be internally cooled during the expansion process.

Conclusions.   The MALDI technique has been most successful for peptides and proteins, which are easily ionized, and do not absorb UV-laser radiation with the same efficiency as the matrix. Thus, applications to the structural analysis of proteins, peptides, and their conjugates are discussed extensively in Chapter 10. Oligonucle-otides have been considerably more difficult to analyze, because the nucleic acid bases in both DNA and RNA absorb strongly at the same wavelengths as most of the common matrices. However, considerable progress has been made in this area and is described in Chapter 11. Carbohydrates and neutral polymers (such as poly-ethylene glycols) are also problematic. While they do not absorb UV radiation, they are not easily ionized; and addition of alkaline salts has often been utilized to enhance the formation of cationized species. MALDI, however, appears to provide the most promising approach for a wide range of biomolecules, with much of the current research devoted to establishing the most suitable matrices for different compound classes.

References

1. Conzemius, R.J.; Capellen, J.M. *Int. J. Mass Spectrom. Ion and Phys.* **34** (1980) 197.
2. Berkowitz, J.; Chupka, W.A. *J. Chem. Phys.* **40** (1964) 2735.
3. Lincoln, K.A. *Anal. Chem.* **37** (1965) 541.
4. Knox, B.E.; Vastola, F.J. *Laser Focus* **3** (1967) 15.
5. Vastola, F.J.; Pirone, A.J.; Giver, P.H.; Dutcher, R.R., in *Spectrometry of Fuels*, Friedel, R.A. Ed.; Plenum Press: New York, NY, 1970; pp. 29–36.
6. Fenner, N.C.; Daly, N.R. *J. Mater. Sci.* **3** (1968) 259.
7. Eloy, J.F. *Methodes Phys. Anal.* **5** (1968) 161.
8. Wiley, R.H.; Veeravagu, P. *J. Phys. Chem.* **72** (1968) 2417.
9. Posthumus, M.A.; Kistemaker, P.G.; Meuzelaar, H.L.C.; deBrauw, M.C. Ten Noever *Anal. Chem.* **50** (1978) 985.
10. Hillenkamp, F.; Unsold, E.; Kaufmann, R.; Nitsche, R. *Nature* **256** (1975) 119.
11. Nitsche, R.; Kaufmann, R.; Hillenkamp, E.; Unsold. E.; Vogt, H.; Wechsung, R. *Isr. J. Chem.* **17** (1978) 181.
12. Krause, M.O.; Vestal, M.L.; Johnston, W.H.; Carlson, T.A., *Phys. Rev.* **133** (1964) A385.
13. Cotter, R.J., *Anal. Chem.* **56** (1984) 485A.
14. Li, L.; Lubman, D.M. *Anal. Chem.* **60** (1988) 2591.
15. Behrson, R.; Even, U.; Jortner, J. *J. Chem. Phys.* **80** (1984) 1050.
16. Park, Y.D.; Rizzo, T.R.; Peteanu, L.A.; Levy, D.H. *J. Chem. Phys.* **84** (1986) 6539.
17. Tembreull, R.; Lubman, D.M. *Anal. Chem.* **59** (1987) 1082.
18. Pellin, M.J.; Young, C.E.; Gruen, D.M., *Scanning Microscopy* **2** (1988) 1353.
19. Li, L.; Lubman, D.M., *Anal. Chem.* **60** (1988) 1409.
20. Lindner, J.; Grotemeyer, J., *J. Molec. Structure* **249** (1991) 81.
21. Grotemeyer, J.; Schlag, E.W., *Org. Mass Spectrom.* **22** (1987) 758.

22. Vastola, F.J.; Mumma, R.O.; Pirone, A.J., *Org. Mass Spectrom.* **3** (1970) 101.
23. Inghram, M.G.; Gomer, R., *Z. Naturforsch.* **10a** (1955) 863.
24. Beckey, H.D., Schuelte, D., *Z. Naturforsch.* **68** (1960) 302.
25. Baldwin, M.A.; McLafferty, F.W., *Org. Mass Spectrom.* **7** (1973) 1353.
26. Ohashi, M.; Yamada, S.; Kudo, H.; Nakayama, N., *Biomed. Mass Spectrom.* **5** (1978) 579.
27. Lee, T.D.; Anderson, W.R., Jr.; Daves, G.D., Jr., *Int. J. Mass Spectrom. Ion and Phys.* **24** (1977) 304.
28. Hunt, D.R.; Shabanowitz, J.; Botz, F.K.; Brent, D.A., *Anal. Chem.* **49** (1977) 1160.
29. Soltmann, B.; Sweeley, C.C.; Holland, J.F., *Anal. Chem.* **49** (1977) 1164.
30. Hansen, G.; Munson, B., *Anal. Chem.* **50** (1978) 1130.
31. Cotter, R.J.; Fenselau, C., *Biomed. Mass Spectrom.* **6** (1979) 287.
32. Reed, R.I., in *Mass Spectrometry*, Reed, R.I., Ed.; Academic Press, London (1965) p. 401.
33. Buehler, R.J.; Flanagan, E.; Greene, L.J.; Friedman, L., *J. Am. Chem. Soc.* **96** (1974) 3390.
34. Cotter, R.J.; Yergey, A.L., *J. Am. Chem. Soc.* **103** (1981) 1596.
35. Cotter, R.J.; Yergey, A.L., *Anal. Chem.* **53** (1981) 1306.
36. Giessman, U.; Rollgen, F.W., *Org. Mass Spectrom.* **11** (1976) 1094.
37. Rollgen, F.W.; Giessman, U.; Heinen, H.J.; Reddy, S., *Int. J. Mass Spectrom. Ion and Phys.* **24** (1977) 235.
38. Heresch, F.; Schmid, E.R.; Huber, J.F.K., *Anal. Chem.* **52** (1980) 1803.
39. Cotter, R.J., *Anal. Chem.* **52** (1980) 1767.
40. Blakley, C.R.; Carmody, J.J.; Vestal, M.L., *J. Am. Chem. Soc.* **102** (1980) 3931.
41. Stoll, R.; Rollgen, F.W., *Org. Mass Spectrom.* **14** (1979) 642.
42. van der Peyl, G.J. Q.; Isa, K.; Haverkamp, J.; Kistemaker, P.G., *Org. Mass Spectrom.* **16** (1981) 416.
43. van der Peyl, G.J. Q.; Haverkamp, J.; Kistemaker, P.G., *Int. J. Mass Spectrom. Ion and Phys.* **42** (1982) 125.
44. Heinen, H.J., *Int. J. Mass Spectrom. Ion and Phys.* **38** (1981) 309.
45. Zackett, D.; Schoen, A.E.; Hemberger, P.H.; Cooks, R.G., *J. Am. Chem. Soc.* **103** (1981) 1295.
46. Graham, S.W.; Dowd, P.; Hercules, D.M., *Anal. Chem.* **54** (1982) 649.
47. Van Breemen, R.B.; Snow, M.; Cotter, R.J., *Int. J. Mass Spectrom. Ion and Phys.* **49** (1983) 35–50.
48. Tabet, J.-C.; Cotter, R.J., *Int. J. Mass Spectrom. Ion Processes* **54** (1983) 151–158.
49. Tabet, J.-C.; Cotter, R.J., *Anal. Chem.* **56** (1984) 1662–1667.
50. Spengler, B.; Dolce, J.W.; Cotter, R.J., *Anal. Chem.* **62** (1990) 1731–1737.
51. Cotter, R.J.; Honovich, J.P.; Olthoff, J.K.; Lattimer, R.P., *Macromolecules* **19** (1986) 2996–3001.
52. Karas, M.; Bachmann, D.; Hillenkamp, F., *Anal. Chem.* **57** (1985) 2935.
53. Tanaka, K.; Ido, Y.; Akita, S., in *Proceedings of the Second Japan-China Joint Symposium on Mass Spectrometry*, Matsuda, H.; Liang, X.-T., Eds.; Bando Press, Osaka (1987) pp. 185–188.
54. Karas, M.; Bachmann, D.; Bahr, U.; Hillenkamp, F., *Int. J. Mass Spectrom. Ion Porcesses* **78** (1987) 53–68.
55. Tanaka, K.; Waki, H.; Ido, Y.; Akita, S.; Yoshida, Y; Yoshida, T., *Rapid Commun. Mass Spectrom.* **2** (1988) 151–153.
56. Karas, M.; Hillenkamp, F., *Anal. Chem.* **60** (1988) 2299–2301.
57. Karas, M.; Ingendoh, A.; Bahr, U.; Hillenkamp, F., *Biomed. Environ. Mass Spectrom.* **18** (1989) 841.
58. Karas, M.; Bahr, U.; Hillenkamp, F., *Int. J. Mass Spectrom. Ion Processes* **92** (1989) 231.
59. Beavis, R.C.; Chait, B.T., *Rapid Commun. Mass Spectrom.* **3** (1989) 233.
60. Salehpour, M.; Perera, I.; Kjellberg, J.; Hedin, A.; Islamian, M.A.; Hakansson, P; Sundqvist, B.U.R., *Rapid Commun. Mass Spectrom.* **3** (1989) 259–263.
61. Spengler, B.; Cotter, R.J., *Anal. Chem.* **62** (1990) 793–796.

62. Wiley, W.C.; McLaren, I.H., *Rev. Sci. Instr.* **26** (1955) 1150.
63. Beavis, R.C.; Chait, B.T., *Rapid Commun. Mass Spectrom.* **3** (1989) 432–435.
64. Nelson, R.W.; Thomas, R.M.; Williams, P., *Rapid Commun. Mass Spectrom.* **4** (1990) 348.
65. Overberg, A.; Karas, M.; Bahr, U.; Kaufmann, R.; Hillenkamp, F., *Rapid Commun. Mass Spectrom.* **4** (1990) 293.
66. Overberg, A.; Karas, M.; Hillenkamp, F., *Rapid Commun. Mass Spectrom.* **5** (1991) 128.
67. Beavis, R.C.; Chaudhary, T.; Chait, B.T., *Org. Mass Spectrom.* **27** (1992) 156–158.
68. Tang, K.; Taranenko, N.I.; Allman, S.L.; Chen, C.H.; Chang, L.Y.; Jacobson, K.B., *Rapid Commun. Mass Spectrom.* **8** (1994) 673–677.
69. Tang, K.; Taranenko, N.I.; Allman, S.L.; Chang, L.Y.; Chen, C.H., *Rapid Commun. Mass Spectrom.* **8** (1994) 727–730.
70. Tang, K.; Allman, S.L.; Jones, R.B.; Chen, C.H., *Org. Mass Spectrom.* **27** (1992) 1389–1392.
71. Michalak, L.; Fisher, K.J.; Alderdice, D.S.; Jardine, D.R.; Willett, G.D., *Org. Mass Spectrom.* **29** (1994) 512–515.
72. Chan, T.-W.D.; Colburn, A.W.; Derrick, P.J., *Org. Mass Spectrom.* **27** (1992) 53–56.
73. Fitzgerald, M.C.; Parr, G.R.; Smith, L.M., *Anal. Chem.* **65** (1993) 3204–3211.
74. Chevrier, M.R.; Cotter, R.J., *Rapid Commun. Mass Spectrom.* **5** (1991) 611–617.
75. Hillenkamp, F.; Karas, M.; Beavis, R.C.; Chait, B.T., *Anal. Chem.* **63** (1991) 1193A–1202A.
76. Zare, R.N.; Levine, R.D., *Chem. Phys. Letters* **136** (1987) 593.
77. Vertes, A.; Gijbels, R.; Levine, R.D., *Rapid Commun. Mass Spectrom.* **4** (1990) 228–233.
78. Vertes, A.; Irinyi, G.; Gijbels, R., *Anal. Chem.* **65** (1993) 2389.
79. Beavis, R.C.; Chait, B.T., *Chem. Phys. Lett.* **5** (1991) 479–484.
80. Huth-Fehre, T.; Becker, C.H., *Rapid Commun. Mass Spectrom.* **5** (1991) 378.
81. Pan, Y.; Cotter, R.J., *Org. Mass Spectrom.* **27** (1992) 3–8.

7

Pulsed Extraction, Continuous Ionization, and Ion Storage Instruments

Matrix-assisted laser desorption/ionization (MALDI) and electrospray ionization (ESI) have greatly extended the kinds of structural problems that can be addressed by mass spectrometry, particularly those involving biological macromolecules. As a pulsed technique, MALDI is easily compatible with time-of-flight mass spectrometry and has been responsible (more than any other technique) for the renewed interest and active development of this mass analyzer. The short pulsed (3 ns to 600 ps) lasers that are currently available obviate the need for the complex pulsed-extraction schemes utilized by Wiley and McLaren,[1] and have resulted in simple, high-voltage extraction instruments that record all of the ions produced from each laser pulse. This same multichannel recording advantage can, of course, be realized on Fourier transform (FTMS) and ion trap (ITMS) mass spectrometers, to which MALDI has also been successfully interfaced.[2,3] However, because of their simplicity, low cost, and high sensitivity, time-of-flight mass spectrometers have enjoyed the most commercial success to date among instruments incorporating MALDI ionization.

In contrast, electrospray ionization is a continuous ionization technique that (like electron impact, chemical ionization, and fast-atom bombardment) is more compatible with scanning instruments, that is, quadrupole and sector (double-focusing) mass spectrometers. In this technique, multiply charged ions are produced from a solution (Figure 7.1) sprayed from a capillary electrode placed at high voltage with respect to a grounded counter electrode. The resultant droplets are dried by a countercurrent flow of a drying gas, and the ions enter the mass spectrometer after passing through a skimmer or capillary tube. While ESI has been utilized on sector instruments, the high gas load, the need for floating the ESI source at high accelerating potential, and the broad range of kinetic energies resulting from collisions in the source have made it a less than ideal configuration for an instrument which (like the time-of-flight mass spectrometer) requires separation of ions accelerated to constant energy. In contrast, quadrupole mass analyzers use grounded ion sources and (as true *massenfilters*) are less affected by the energy spread. While quadrupole mass analyzers

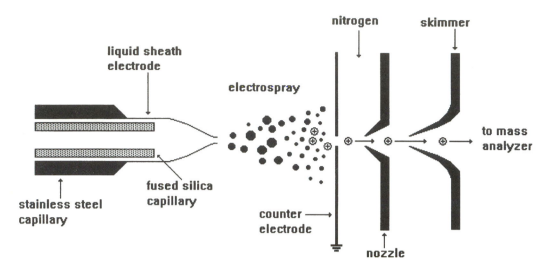

FIGURE 7.1   Diagram of an electrospray ionization source.

generally have mass ranges of only 2000 Da, the production of multiply charged ions extends the effective mass range into the hundreds of kilodaltons. In addition, the internal instability of such ions (due to charge repulsion) provides an opportunity to achieve efficient fragmentation using low-energy collision-induced dissociation (CID). Thus, the most effective (and commercially successful) instrumental combination uses an ESI source on a triple quadrupole mass spectrometer.

The ability to provide online LC–MS and LC–MS–MS is perhaps the most attractive feature of electrospray ionization, and has motivated its use on other mass analyzers, including the ion trap[4] and the Fourier transform mass spectrometer.[5] It has also motivated development of ESI–TOF configurations, despite the fact that ESI is continuous and produces ions with a high kinetic energy distribution. Thus, in this chapter we consider the general problem of utilizing continuous ionization sources on a time-of-flight mass spectrometer and the use of beam deflection and pulsed extraction to provide an initial ion packet, orthogonal extraction to minimize the effects of initial kinetic energy distribution, and ion storage techniques to improve the duty cycle. While these methods are focused primarily on the development of ESI–TOF configurations, we also consider the use of delayed, pulsed ion extraction for laser desorption and MALDI techniques to improve both mass resolution and structural analysis through fragmentation in the ion source.

Continuous Ionization Sources

The first time-of-flight mass spectrometers used electron impact (EI) ionization, so that the need for accomodating continuous ionization sources was recognized from the beginning. Many of these early instruments utilized beam-deflection techniques[6–9] to admit a narrow initial ion packet into the flight tube. Later, Pinkston et al.[10] utilized beam deflection in an E-TOF (where E is an electrostatic energy

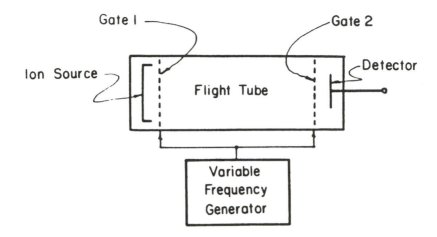

FIGURE 7.2 Schematic diagram of a Fourier transform time-of-flight mass spectrometer. (Reprinted with permission from reference 13).

analyzer) and noted that beam deflection avoids the problem associated with turn-around time. Beam deflection techniques have also been utilized on time-of-flight mass spectrometers using an atmospheric pressure ionization (API) source,[11] and a variation on the beam deflection method, known as beam voltage modulation, has been proposed for continuous sources.[12]

Pulsed Extraction Techniques.   However, the most common approach, developed by Wiley and McLaren,[1] addressed the problem of continuous ionization (in an EI source) by pulsing both the ionizing electron beam and the ion extraction voltage. While we have emphasized the spatial and kinetic energy focusing capabilities of their time-lag focusing scheme (Chapter 2), it should be noted that the 1- to 10-μs electron-beam pulse produced ionization periods that (for purposes of mass resolution) can be regarded as continuous. Thus, the primary function of their delayed, pulsed extraction technique was the rapid ejection of ions from the source by a fast (40-ns rise time) ion extraction pulse, that is, time focusing. As detailed below, similar schemes have been utilized to reduce the initial ion-packet width of ions produced by broad laser pulses or energetic ion beams.

Fourier Transform TOF.   A novel approach to the problem of continuous ionization that also attempts to improve the duty cycle was proposed by Fritz Knorr and Dale Chatfield.[13] In their scheme (Figure 7.2), a symmetric square waveform is applied synchronously to the entrance (drawout) and exit grids bounding the flight tube. Because the waveform is symmetric, one-half of the ions are passed by each grid at all frequencies, resulting in an overall transmission of 25% of the ions. However, the ions that are able to successfully pass through both grids and reach the detector must have the correct velocity (i.e., mass if they are accelerated to constant energy). Thus, a mass spectrum is recorded by modulating the frequency, and transforming the ion signal obtained in the time domain to the frequency domain.

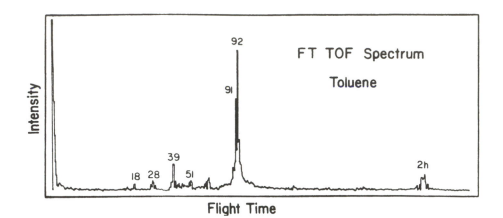

FIGURE 7.3   Fourier transform time-of-flight mass spectrum of toluene. (Reprinted with permission from reference 13).

An example of a Fourier transform TOF mass spectrum of toluene obtained from this instrument is shown in Figure 7.3. Their overall duty cycle of 25% is considerably greater than that of the Wiley-McLaren design for electron impact ionization.

Combining Delayed and Pulsed Ion Extraction with Desorption Techniques

Although lasers with pulse widths in the range of 3 ns to 600 ps are commonly used in MALDI time-of-flight instruments, such short pulse widths have not always been available. Thus, beginning in 1980, our laboratory developed a series of delayed-extraction instruments using infrared laser desorption (IRLD), liquid SIMS, and (ultimately) MALDI. The IRLD–TOF instrument shown in Figure 7.4 (along with its timing diagram) utilized a carbon dioxide laser with a wavelength of 10.6 μm.[14,15] Carbon dioxide lasers (typically) have initial high-energy pulse widths of about 40–50 ns, followed by a longer (up to 100 μs) low-energy tail. In addition, the extended thermal emission of ions occuring for many microseconds following the rather large energy (10–100 mJ) pulse was (as far as time focusing is concerned) a continuous source. Thus, pulsed ion extraction provided a means to obtain a well-focused initial ion packet. While implemented on a Wiley-McLaren mass spectrometer, the time-delayed extraction approach used in this instrument was distinctly different from the time-lag focusing scheme. Using delay times of up to 200 μs, pulsed extraction was used to probe the contents of the source at different times after the laser pulse. This promoted the formation of ions by gas-phase processes (primarily attachment of alkali ions) and allowed one to observe structurally significant fragmentation in the mass spectrum, occuring in the same time frame as generally observed in post-source decay spectra (Chapter 8) using prompt extraction instruments. Figure 7.5 shows an example of a an IRLD mass spectrum of the lipid A from *Rhodopseudomonas sphaeroides,* obtained by extracting the ions more than 20 μs after the laser pulse. Molecular-ion species are formed by attachment of K+ ions, and abundant fragmentation is observed.

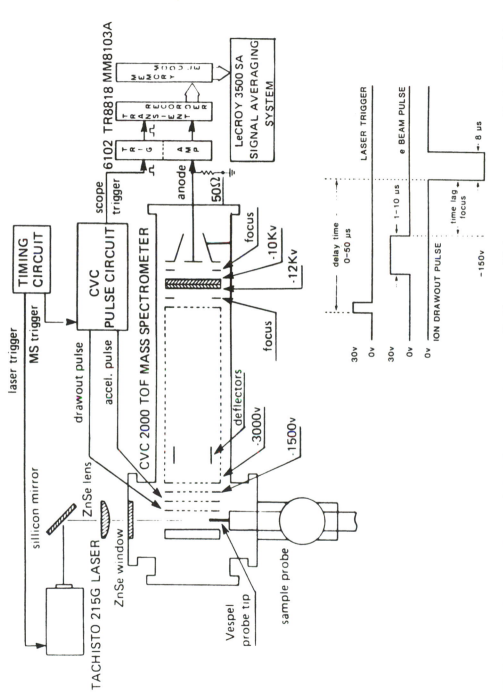

FIGURE 7.4 (a) Schematic diagram of the infrared laser desorption (IRLD) time-of-flight mass spectrometer with time-delayed ion extraction, and (b) timing diagram. (Reprinted with permission from reference 15).

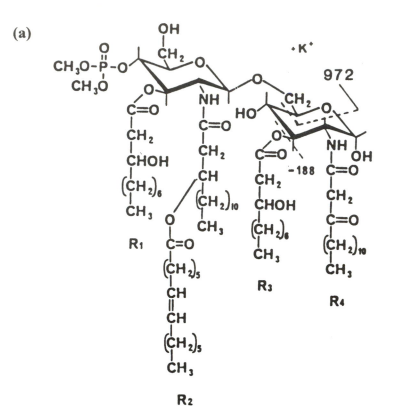

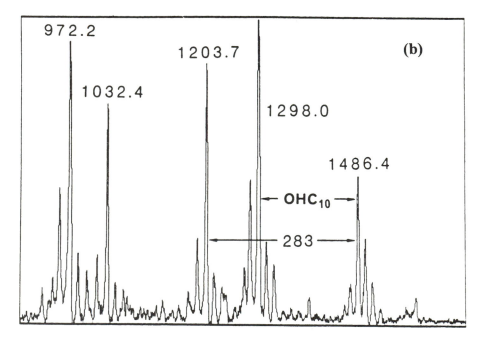

FIGURE 7.5   Structure and delayed-extraction infrared laser desorption (IRLD) mass spectrum of the lipid A from the bacterium *R. sphaeroides*. (Reprinted with permission from reference 15).

Passive Ion Storage. The IRLD instrument was constructed from a modified CVC Products (Rochester, NY) model 2000 time-of-flight mass spectrometer, a Wiley-McLaren design that included a pulsed electron beam. When used in the electron impact (EI) ionization mode it was possible to monitor the N_2^+ ion signal as a function of the electron-beam pulse width from 0.1 μs (where the signal was proportional to the pulse width) to 3 μs (where losses due to ion drift from the source equal the rate of ion production). From this, one could determine the rate of ion loss from the extraction volume due to drift at thermal energies. For the N_2^+ ion ($m/z = 28$) the half-life in the extraction volume was about 1 μs.[16] Since the drift velocity is:

$$v = \left(\frac{2U_0}{m} \right)^{1/2} \qquad [1]$$

larger ions with similar thermal energies (U_0) will have half-lives in the extraction region that vary with the square root of their mass. For example, an ion of m/z 2800, formed at thermal energies, should have a half-life of about 10 μs. Thus, prior to the application of an extraction field, the field-free ion source in a Wiley-McLaren instrument provides a kind of passive ion storage device (i.e., no electrical fields are used to confine the ions), whose storage time improves with mass.

It was this passive ion storage capability that enabled the construction of a liquid SIMS (secondary-ion mass spectrometry) time-of-flight mass spectrometer intended to provide results similar to those observed by fast-atom bombardment.[17] You may recall (from Chapter 5) that static SIMS–TOF instruments utilize a short (1 to 2 ns) burst of energetic 1-nA/cm², primary-ion beams, while fast-atom bombardment utilizes continuous primary beams of the order of 1 μA/cm² and a liquid matrix. In the liquid SIMS–TOF instrument described in Chapter 5, high flux (1–10 μA/cm²), broad pulse (1–10 μs) primary-ion beams were utilized, with time focusing of ions within the extraction volume provided by pulsed, delayed extraction. Like the IRLD instrument described above, ion extraction was orthogonal to the desorption direction. Allowing the ions to drift into the extraction region at right angles to the time-of-flight axis provided the best means for focusing ions desorbed from an irregular, nonconducting surface containing the liquid matrix.

Energy Spread. Orthogonal and delayed ion extraction was also utilized on the MALDI–TOF instrument shown in Figure 7.6.[18] In this case, the objective was to mitigate the effects of the high energy spread of MALDI ions on mass resolution. As shown in the detailed diagram of the source region (Figure 7.7), the time delay could be varied to sample ions within a given velocity range. This instrument was, in fact, utilized to map the velocity and energy distributions of MALDI ions.[19]

Fragmentation in the Ion Source. A major reason that time-of-flight (generally PDMS and MALDI) mass spectrometers have come to be regarded as molecular weight instruments is that the ions are extracted promptly, before fragmentation can occur; hence, the current interest in post-source decay methods (Chapter 8). Delayed ion extraction provides an opportunity to carry out fragmentation in the ion source. For example, Figure 7.8 compares the PDMS spectrum of a phospholipid (using

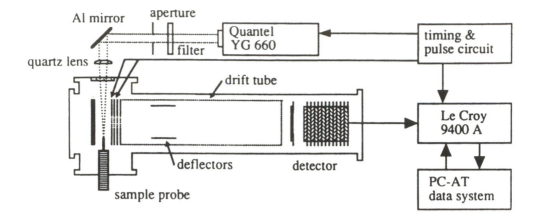

FIGURE 7.6 Schematic of a delayed-extraction MALDI–TOF mass spectrometer. (Reprinted with permission from reference 18).

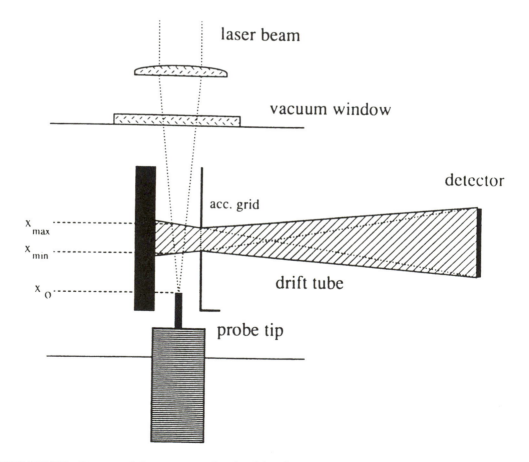

FIGURE 7.7 Diagram of the ion source for the delayed-extraction MALDI–TOF mass spectrometer. Ions within x_{max} and x_{min} are transmitted to the detector. (Reprinted with permission from reference 18).

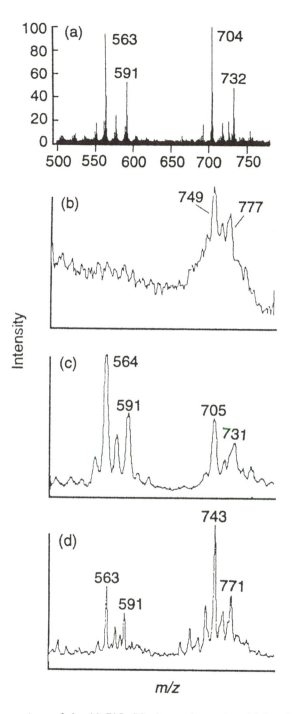

FIGURE 7.8 Comparison of the (a) FAB, (b) plasma desorption, (c) liquid SIMS, and (d) IRLD mass spectra of the phospholipids extracted from *E. coli*. The mass spectra in (c) and (d) were obtained by delayed extraction after 20 and 50 μs, respectively. The molecular ions are MH$^+$ in the FAB (a) and liquid SIMS spectra (c), M-H+2Na$^+$ in the plasma desorption mass spectrum, (b) and MK$^+$ in the IRLD mass spectrum (d). (Reprinted with permission from reference 20).

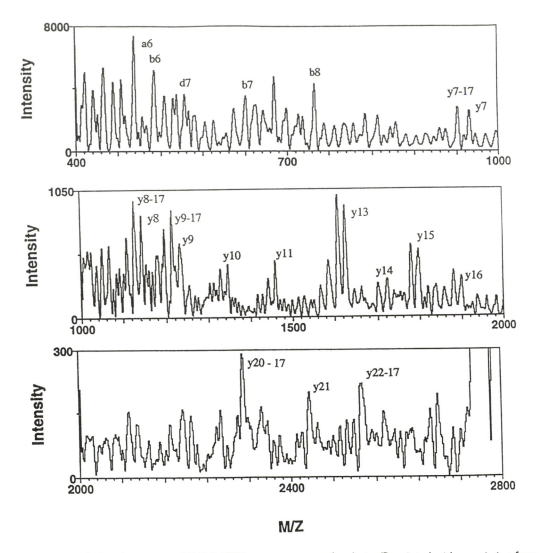

FIGURE 7.9   Delayed extraction MALDI–TOF mass spectrum of melittin. (Reprinted with permission from reference 21).

prompt extraction) with that obtained by delayed extraction IRLD and liquid SIMS, and the FAB spectrum obtained on a sector instrument.[20] In the PDMS spectrum, the metastable fragmentation occuring in the ion source during prompt extraction results in considerable peak broadening. In the delayed extraction IRLD and liquid SIMS spectra, the fragment ions are resolved, and the mass spectrum resembles that obtained on a sector instrument.

More recently, delayed orthogonal extraction has been utilized to observe peptide fragmentation on MALDI time-of-flight instruments. Figure 7.9 shows the high irradiance MALDI mass spectrum of melittin obtained by Standing et al.,[21] following an 18-μs delay.

Orthogonal Extraction

The advantages of orthogonal extraction received considerable attention following reports by Dawson and Guilhaus[22] in 1989 and Dodenov et al.[23] in 1991. In the Dawson and Guilhaus report, orthogonal extraction was used to improve both the duty cycle and mass resolution for a linear, electron impact (EI) ionization mass spectrometer. In the Dodenov report, orthogonal extraction was used on a reflectron instrument to provide compatibility of the TOF mass analyzer with the continuous, high-energy-spread ions formed by electrospray ionization. As will be shown later in Chapter 9, interest in orthogonal extraction also now includes its use in a hybrid tandem (EB–TOF) mass spectrometer.

Improving the Duty Cycle.   The instrument of Dawson and Guilhaus[22] was developed as an alternative to the time-lag focusing scheme of Wiley-McLaren[1] for focusing in the gas phase (rather than from an equipotential surface). As shown in Figure 7.10, orthogonal extraction serves two functions. First, the collimating lens converts y-axis components of kinetic energy to the x-axis, which is orthogonal to the time-of-flight direction. Thus, upon application of the push-out pulse, ions are extracted into the time-of-flight analyzer that are energy-focused in that direction, while the drift length itself is set at an angle to account for the off-axis drift of the ions. In addition, the combination of voltages and distances for the push-out plate, and the three extraction grids (Figure 7.11) make up a dual extraction system that provides space focusing of the broad orthogonal beam at the detector. The second function, ion storage, operates as the orthogonal accelerator alternates between the *fill-up* and *push-out* modes, and is intended to improve the duty cycle when the instrument is used as a gas-chromatographic detector. In addition, unlike the Wiley-McLaren instrument, the electron beam can be operated continuously, since ions formed during the push-out cycle (as shown in Figure 7.11) will not be extracted into the drift region. Figure 7.12 shows the mass spectrum of tris-2,4,6-(pentadecafluoro-heptyl)-1,3,5-triazine (PFHT) obtained by orthogonal extraction.[24] Mass resolution for the M-F$^+$ ion at m/z 1166 (Figure 7.12b) is better than 1 part in 4000.

Electrospray Ionization on TOF Instruments.   Dodenov and co-workers first reported their design for an electrospray ionization time-of-flight mass spectrometer using an atmospheric pressure ionization (API) source and orthogonal extraction at the 12th International Mass Spectrometry Conference held in 1991 in Amsterdam.[23] Published later as a book chapter,[25] their paper described the instrument shown in Figure 7.13. As in the Guilhaus instrument, the ions are collimated along a direction orthogonal to the time-of-flight axis, in this case using a quadrupole lens (**3** in Figure 7.13b). Ions are extracted from the storage volume by a combination of push-out and draw-out pulses applied to the backing plate and second extraction grid (**7** and **8** in Figure 7.13b), respectively. In addition, the instrument incorporates a dual-stage reflectron, with the off-axis drift controlled by deflection lenses (**4** in Figure 7.13a). This instrument was intended to compensate (by orthogonal extraction) for the large kinetic energy spread of ions produced by electrospray ionization, and to provide compatability (using storage techniques) with this continuous ionization source. Figure 7.14 shows ESI–TOF mass spectra of hen egg-white lysozyme and the pepsin digest of reduced and pyridylethylated porcine neutrophil peptide obtained on their instrument.[25,26]

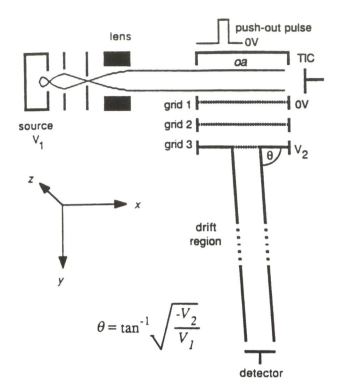

$$\theta = \tan^{-1}\sqrt{\frac{-V_2}{V_1}}$$

FIGURE 7.10   Schematic diagram of the orthogonal-extraction time-of-flight mass spectrometer. (Reprinted with permission from reference 22).

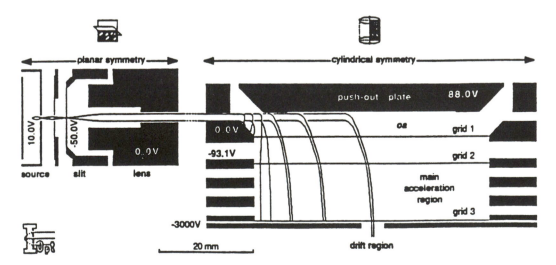

FIGURE 7.11   Details of the ion source and storage region of the orthogonal-extraction time-of-flight mass spectrometer. (Reprinted with permission from reference 22).

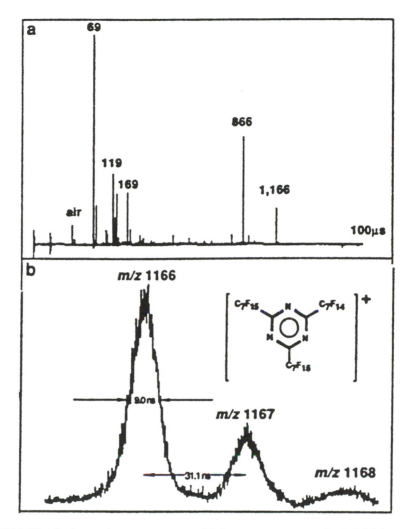

FIGURE 7.12   Orthogonal-extraction time-of-flight mass spectrum of tris-2,4,6-(pentadecaflu-oro-heptyl)-1,3,5-triazine (PFHT). (a) Complete mass spectrum. (b) Expansion of the isotopic peaks for the M-H⁺ ion. (Reprinted with permission from reference 24).

Verentchikov and co-workers[27] incorporated orthogonal injection of ESI ions on a time-of-flight instrument with a single-stage reflectron (Figure 7.15). Figure 7.16 shows their results for the peptide melittin using different nozzle–skimmer voltages in the ESI source to vary the extent of fragmentation.

Combining Orthogonal Extraction with Pulsed Injection.   It is possible to improve the performance of orthogonal extraction TOF instruments by limiting access of ions into the storage region during the extraction cycle. Such an instrument was designed by Hieftje and co-workers for atmospheric pressure ionization (API), inductively coupled plasma (ICP) mass spectrometry using first a cylindrical lens deflection lens system,[28] and later the quadrupole injection lens system[29] shown in

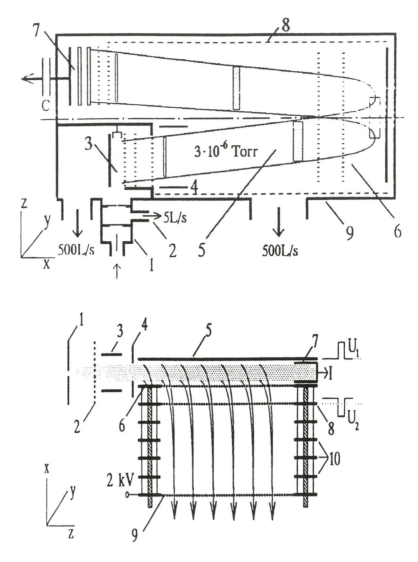

FIGURE 7.13 Schematics of (a) the API–TOF mass spectrometer with an ion mirror and (b) the ion-storage modulator. (Reprinted with permission from reference 25).

Figure 7.17. Modulation of the Q_x poles (where the x-axis is the time-of-flight axis) provides considerably better efficiency than modulation of the Q_y poles, and (as shown in the mass spectra of Pb and Bi in Figure 7.18) better mass resolution. In addition, an Ar deflection pulse provides mass selective removal of Ar^+ ions arising from the argon plasma of the ICP source, and reduces the possibilities for saturation of the microchannelplate detectors.

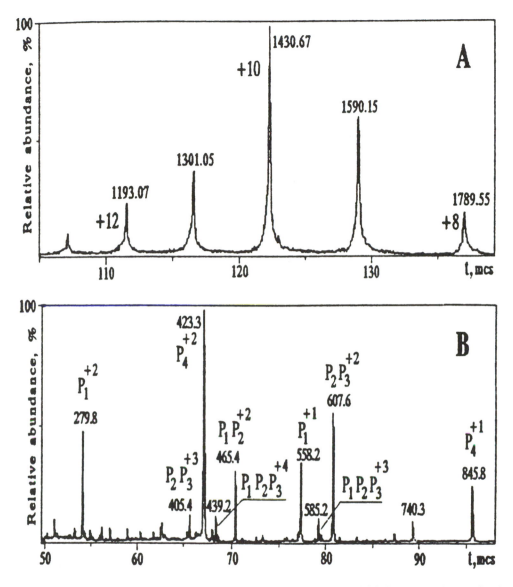

FIGURE 7.14 ESI–TOF mass spectra of (a) hen egg-white lysozyme and (b) the pepsin digest of reduced and pyridylethylated porcine neutrophil peptide. (Reprinted with permission from references 25 and 26).

Other Ion Storage Schemes

Wollnik et al.[30] described a method for storing ions produced by electron impact (EI) ionization that utilizes a series of three grids whose potentials are changed between the ion storage and extraction periods. As shown in Figure 7.19, the potentials on grids G_2 and G_4 provide a potential well at grid G_3 during the ion storage period. In the extraction period, the potential on grid G_3 is raised, resulting in

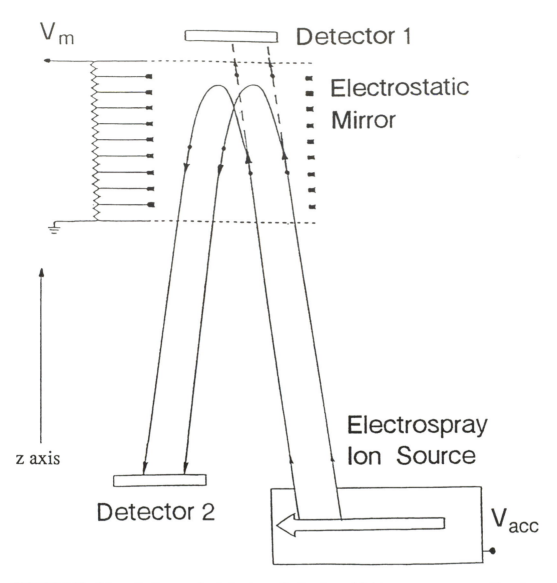

FIGURE 7.15   Schematic diagram of a single-stage reflecting time-of-flight mass spectrometer with orthogonally injected ESI ions. (Reprinted with permission from reference 27).

acceleration of the ions and focusing at the space-focus plane (SP). An electrospray time-of-flight instrument utilizing a similar electrostatic system for ion bunching was developed by Boyle et al.[31] and is shown in Figure 7.20, along with a mass spectrum of cytochrome c. For combining electrospray ionization and time-of-flight mass spectrometry, such bunching methods have not proven to be as successful as orthogonal extraction techniques, or (as described below) ion storage using a combined ion-trap TOF mass spectrometer.

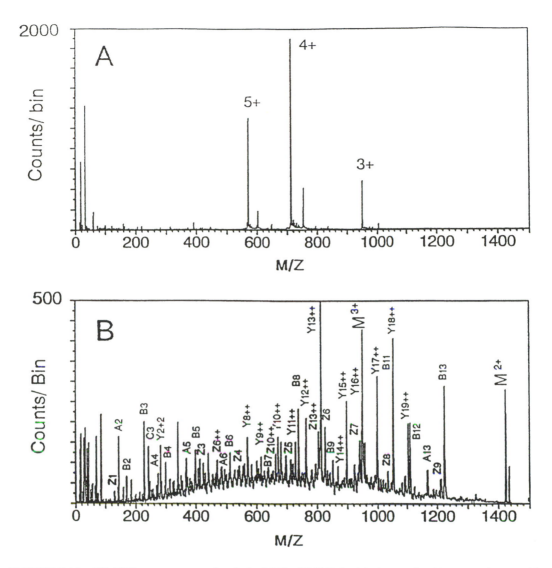

FIGURE 7.16   ESI–TOF mass spectra of melittin (MW = 2846 Da) with the nozzle–skimmer voltage at (a) 100 V and (b) 300 V. (Reprinted with permission from reference 27).

The Ion-Trap Time-of-Flight Mass Spectrometer.   Lubman et al.[32] described a combination ion-trap time-of-flight mass spectrometer that was used initially for recording ions produced by matrix-assisted laser desorption.[33] The idea of combining an ion trap with a time-of-flight spectrometer is an interesting one. As an ion storage device, the ion trap provides a suitable interface between a continuous ionization source and the time-of-flight mass analyzer, and thus motivated Lubman's development of an electrospray-ionization ion-trap TOF instrument,[34] in which the ESI ion beam is inline with the time-of-flight axis. The ion trap also provides the opportunity

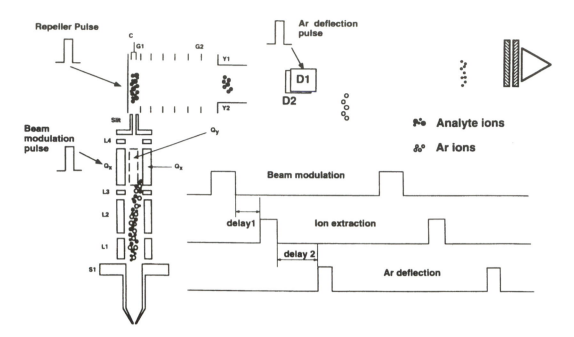

FIGURE 7.17 Schematic diagram of the pulsed ion-injection process showing the beam-modulation pulse applied to the quadrupole lens and its timing relationship with the orthogonal extraction (repeller) and Ar deflection pulses. (Reprinted with permission from reference 29).

for reducing the kinetic energy spread through collisional cooling with the helium trapping gas, and for removal of low-mass (particularly matrix) ions in order to reduce saturation effects on the TOF channelplate detector.

In the instrument designed by Lubman and co-workers (Figure 7.21), electrospray ions are collimated by an Einsel lens as they enter the trapping field defined by the radio frequency voltage (V_{rf}) on the ring electrode and the grounded endcaps. The ions are then injected into the time-of-flight analyzer by application of an extraction pulse (V_{extr}) on the exit endcap, collimated by a second Einsel lens, and steered into the center of an ion reflector. Figure 7.22a shows the ESI–TOF mass spectrum of the peptide Trp-Met-Asp-Phe-amide (MW = 596 Da) with the expanded view of the molecular-ion region revealing a mass resolution of 1500 (FWHM). In addition, they were able to induce fragmentation in the ESI source by increasing the nozzle voltage (V_{lens}), as shown in Figure 7.22b.[34]

Fragmentation can also be carried out in the ion trap. When ionization is carried out by MALDI, considerable metastable fragmentation results from the use of higher laser powers. As will be described in Chapter 8, such fragmentation can be observed in reflectron TOF instruments along a different (usually linear) mass scale. In the ion-trap TOF instrument, the metastable fragmentation induced by MALDI can be observed by increasing the trapping time prior to application of the extraction pulse. The resultant fragment ions acquire the same kinetic energy as their precursors, and are thus observed on the same square-root scale. Figure 7.23 shows the MALDI ion-trap TOF mass spectrum obtained by Lee and Lubman[35] after 10 ms of trapping time.

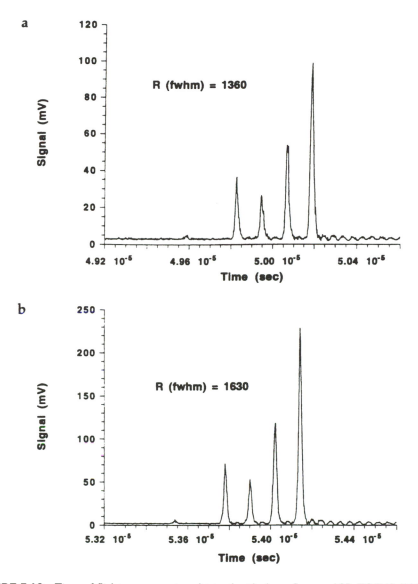

FIGURE 7.18 Time-of-flight mass spectra obtained with the reflectron ICP–TOFMS (1000-shot average) for pulsed ion injection utilizing (a) the Q_y poles (4-ppm solution of Pb and Bi) and (b) Q_x poles (40-ppm solution of Pb and Bi). (Reprinted with permission from reference 29).

Neither of these approaches to fragmentation achieve the tandem (MS–MS) capabilities that should be possible using the ion-trap TOF combination. With the addition of a supplementary rf voltage on the endcaps it should indeed be possible to isolate the molecular ions of a particular analyte in a mixture[36,37] and provide resonant excitation of a given molecular ion[37,38] prior to extraction of the ions into the time-of-flight mass analyzer.

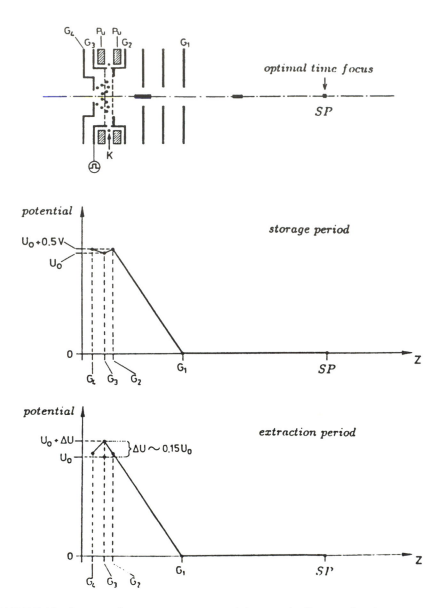

FIGURE 7.19   Storage of ions using a system of three grids. (Reprinted with permission from reference 30).

Revisiting Delayed Extraction for MALDI Ions

Time-lag focusing as described by Wiley and McLaren[1] was intended to provide simultaneous focusing of the initial spatial and kinetic energy distributions for ions formed in the gas phase. In their technique, the amplitude of the extraction (drawout) pulse, in combination with additional constant extraction voltages, made up a dual-stage extraction system that provided mass-independent space focusing at the detector. Energy focusing was provided by the length of the time-delay, and was mass-dependent.

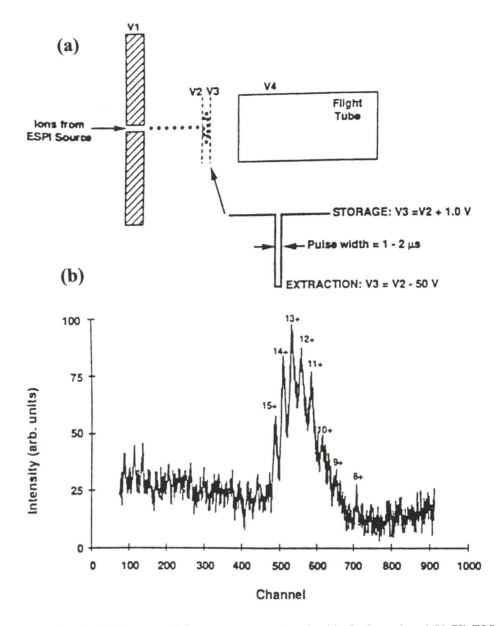

FIGURE 7.20 (a) Schematic of the ion storage optics utilized by Boyle et al., and (b) ESI–TOF mass spectrum of cytochrome c (MW = 12,360.9 Da) using ion storage. (Reprinted with permission from reference 31).

In contrast, the time-delayed extraction used by Cotter and co-workers was intended to improve the mass resolution of ions desorbed from surfaces by infrared-laser desorption,[14] liquid SIMS,[17] and MALDI.[18] In all cases, desorption was orthogonal to the time-of-flight axis, so that focusing effects were applied to the much narrower kinetic energy distribution along the time-of-flight direction. The optimal length of the time delay was not particularly mass dependent. In the case of IRLD

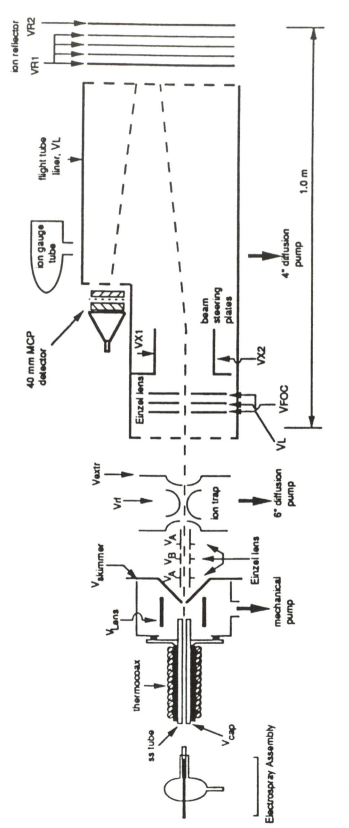

FIGURE 7.21 External ion-injection ion-trap-storage reflectron time-of-flight mass spectrometer with an electrospray source with: V_A, V_B pretrap Einzel lens voltages; V_{rf} radio frequency potential; V_{extr} dc extraction voltage; $V_L = V_{x1}$ flight tube liner voltage; V_{foc} focusing voltage; V_{x2} beam steering voltage; V_{R1} ion repeller voltage; and V_{R2} ion reflectron voltage. (Reprinted with permission from reference 34).

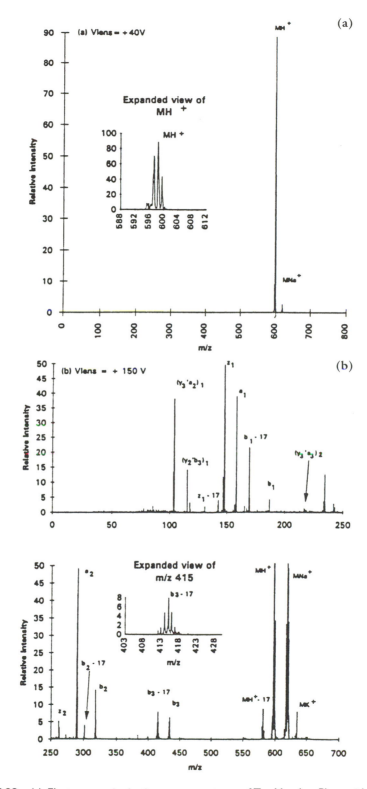

FIGURE 7.22   (a) Electrospray ionization mass spectrum of Trp-Met-Arg-Phe amide with V_{lens} = +40V, and (b) electrospray ionization CID mass spectrum with V_{lens} = +150V. (Reprinted with permission from reference 34).

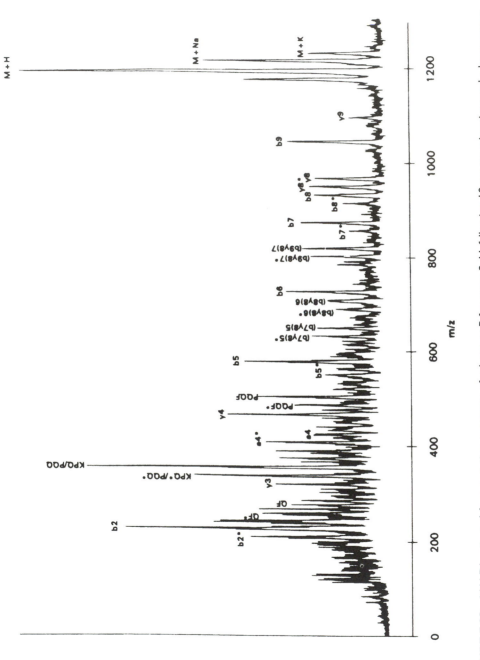

FIGURE 7.23 MALDI-activated fragmentation spectrum of substance P fragment 2-11 following 10-ms trapping time and a laser power density of 3.5×10^6 W/cm^2. (Reprinted with permission from reference 35).

and liquid SIMS, pulsed extraction provided (primarily) a sharper start pulse than provided by the IR laser pulse or a 1- to 10-μs primary-ion beam pulse. For MALDI, the technique took advantage of the fact that ions produced by this technique are entrained in the expanding plume of matrix ions[19,39,40] and drift into the extraction region with velocities that are relatively independent of mass. Additionally, these time-delayed techniques were intended to provide space-focusing of ions produced by gas-phase cationization, and to generate and observe in-source fragmentation (ISD).

Delayed Extraction MALDI.   Recently, time-lag focusing techniques have been applied in MALDI instruments. In these applications, ions are desorbed along the same direction as the time-of-flight axis, from equipotential surfaces (probes) located in the backing plate of the ion source. Thus, the method is clearly similar to that of Wiley and Mclaren, focusing the kinetic energy distribution along the time-of-flight axis. A major difference, however, is that ions formed from an equipotential surface have an initial kinetic energy distribution, but not an initial spatial distribution. During the time delay between ionization and extraction, ions with different initial kinetic energies will distribute themselves in the field-free ion source along the extraction direction, which will (in effect) convert their energy distributions into a spatial distribution (although the kinetic energy distribution is not lost). This is illustrated in Figure 7.24 in which is represented a series of ions with differences in initial kinetic energy (Figure 7.24a), different initial spatial locations that can be focused to a space-focus plane (Figure 7.24b), and the distribution of ions with initial kinetic energy after a time delay (Figure 7.24c). In the latter example, the conditions for focusing of ions will be mass-dependent, but (as described in the reports below) this mass-dependent focusing can be accomplished by changing the amplitude of the extraction pulse rather than the delay time.

For example, Whittal and Li[41] have described a linear, delayed extraction MALDI–TOF instrument which employs a constant (280 ns) delay time while increasing the extraction-pulse potential with mass. Figure 7.25 shows a series of delayed-extraction MALDI mass spectra obtained at their respective optimal pulsed-extraction voltages that show some impressive mass resolutions at high mass. The mass-dependent nature of this approach was also illustrated in a report by Brown and Lennon[42] in which they determined (in Figure 7.26) the optimal extraction-pulse voltage amplitudes for two different delay times and accelerating (source bias) voltages. At the same time, it has been suggested that there are two important effects of delayed extraction that are not particularly mass-dependent. The first arises from the possibility that mass resolution is degraded by collisions of analyte ions with the matrix debris when these ions are extracted in a high electrical field, an idea that was first suggested by Standing et al.[43] Thus, delayed extraction provides an opportunity for dissipation of the matrix plume prior to application of the extraction field. The second is the possibility that delayed extraction provides better temporal focusing, forming an initial ion packet whose time width is shorter than the width of the ionizing laser pulse, or compensates for the formation of ions by subsequent gas-phase, chemical-ionization reactions. In this case (as described in Chapter 2), mass resolution is improved by increasing the flight length. Thus, Vestal et al.[44] have compared the improvements in mass resolution using delayed extraction on instruments having flight-tube lengths of 1.3, 2.0, and 4.2 m. Figure 7.27 shows the mass spectra for the protonated molecular-ion region of bovine insulin (MW = 5733.5

(a) initial kinetic energy distribution

(b) initial spatial distribution

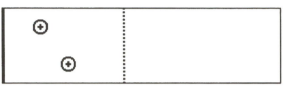

(c) initial energy distribution after time delay

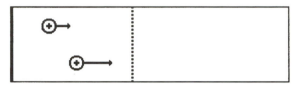

FIGURE 7.24   Schematic comparison of the relative energies and positions of ions having (a) an initial kinetic energy, (b) an initial spatial distribution, and (c) an initial kinetic energy after a time delay.

Da) obtained in the reflectron mode (corresponding to effective path lengths of 2.0, 3.0, and 6.6 m, respectively). In the longest (4.2 m) instrument, mass resolution for the MH+ ion cluster is 1 part in 12,500, although the peak corresponding to the MH+-H$_2$O ions remains unresolved.

In-Source Decay (ISD).   In the next chapter we will describe a method for amino acid sequencing of peptides on reflectron TOF instruments known as post-source decay (PSD). This method exploits the considerable metastable fragmentation that occurs in the drift region after the ions are extracted from the ion source. The resulting product ions spend a shorter time in the reflectron, and therefore arrive at the detector sooner than their precursors.

When delayed extraction is used, metastable fragmentation in the same time frame now occurs in the ion source, and will be observed in the mass spectra obtained from both linear and reflectron instruments. Figure 7.28 shows the in-source decay MALDI mass spectrum of the oxidized B chain of bovine insulin obtained using a 1.5-kV extraction pulse applied 350 ns after the laser pulse.[45] While ISD has only recently been exploited for the purpose of observing fragmentation in MALDI mass spectra, the promotion of in-source fragmentation by delayed extraction has been a feature of instruments using other ionization methods. Examples already described in this book include the delayed-extraction IRLD mass spectra of peptides and oligosaccharides (Figures 6.12 and 6.13 in Chapter 6) and the delayed extraction IRLD and liquid SIMS spectra of phospholipids (Figure 7.8).

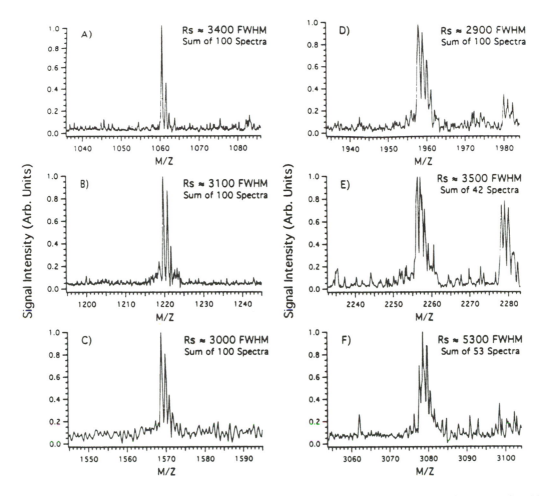

FIGURE 7.25 Delayed-extraction MALDI mass spectra of the molecular ion regions of the peptides: (a) bradykinin (MH⁺, MW = 1060.6) obtained at an extraction pulse potential of 0.9 kV, (b) *LYPVKLPVK* (MH⁺, MW = 1219.7) at 1.0 kV, (c) Ac-*KLEALEAKLEALEA*-NH₂ (MH⁺, MW = 1568.3) at 1.3 kV, (d) Ac-*EAEKAAKE-AEKGAKEAEK*-NH₂ (MH⁺, MW = 1958.0) at 1.75 kV, (e) Ac-*ELEKLLKECEKLLKELEK*-NH₂ (MH⁺, MW = 2256.3) at 2.05 kV, and (f) Ac-*KLEALEAKLEALEAKLEALEAKLEALEA*-NH₂ (MH⁺, MW = 3077.7) at 2.5 kV. (Reprinted with permission from reference 41).

The use of delayed extraction to provide in-source fragmentation has some distinct advantages over the post-source decay method that is described in the next chapter. When molecular and fragment ions are all formed in the source, they are all accelerated to the same kinetic energy, optimally focused by the reflectron, and appear along the same square-root mass scale. In contrast, fragment ions formed by post-source decay have different kinetic energies, follow a different mass scale than precursor ions, and are generally focused by stepping the reflectron voltage. At the same time the ISD approach does not provide the opportunity for precursor mass selection when mixtures are involved.

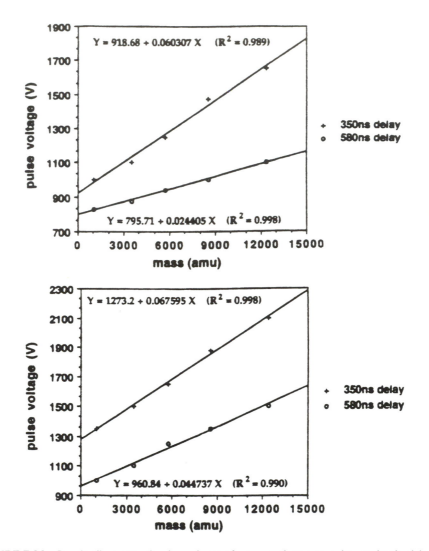

FIGURE 7.26 Results illustrating the dependence of mass resolution on pulse amplitude, delay time, and source bias voltages of (a) 15 kV and (b) 24 kV. (Reprinted with permission from reference 42).

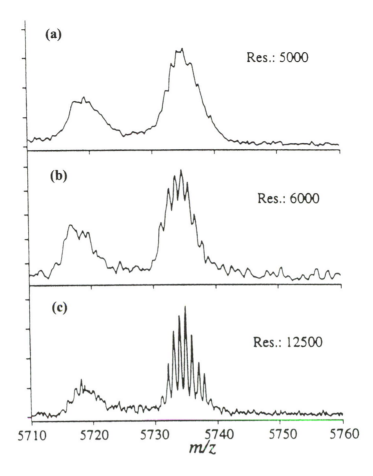

FIGURE 7.27 Delayed-extraction MALDI mass spectra of the molecular-ion region of bovine insulin (MW 5733.4) obtained on three separate reflectron TOF instruments having effective path lengths of (a) 2.0 m, (b) 3.0 m, and (c) 6.6 m. (Reprinted with permission from reference 44).

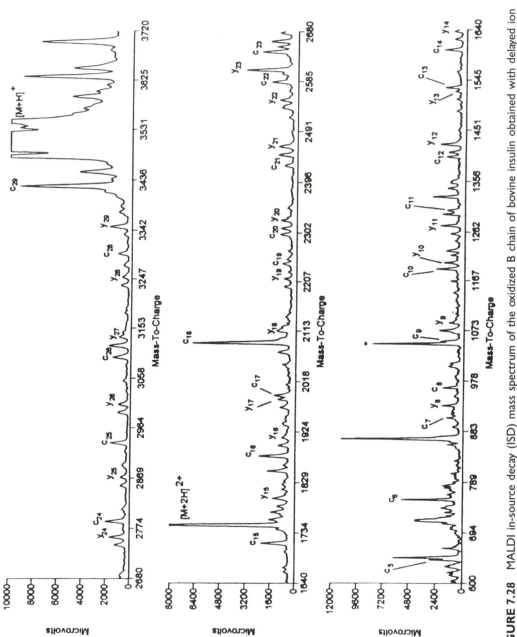

FIGURE 7.28 MALDI in-source decay (ISD) mass spectrum of the oxidized B chain of bovine insulin obtained with delayed ion extraction. The extraction pulse was 1.5 kV applied 350 ns after the laser pulse. (Reprinted with permission from reference 45).

References

1. Wiley, W.C.; McLaren, I.H., *Rev. Sci. Instr.* **26** (1955) 1150–1157.
2. Castoro, J.A.; Koster, C.; Wilkins, C., *Rapid Commun. Mass Spectrom.* **6** (1992) 239.
3. Doroshenko, V.M.; Cornish, T.J.; Cotter, R.J., *Rapid Commun. Mass Spectrom.* **6** (1992) 753–757.
4. Van Berkel, G.J.; Glish, G.L.; McLuckey, S.A., *Anal. Chem.* **62** (1990) 1284.
5. Hofstadler, S.A.; Laude, D.A., *Anal. Chem.* **64** (1992) 572–575.
6. Cameron, E.; Eggers, D.F., Jr., *Rev. Sci. Instr.* **19** (1948) 605.
7. Fowler, T.K.; Good, W.M., *Nucl. Instrum. Meth.* **7** (1960) 245.
8. Bakker, J.M., *J. Phys.* **E6** (1973) 785.
9. Bakker, J.M., *J. Phys.* **E7** (1974) 364.
10. Pinkston, J.D.; Rabb, M.; Watson, J.T.; Allison, J., *Rev. Sci. Instr.* **57** (1985) 583.
11. Ma, C.; Michael, S.M.; Chien, M.; Zhu, J.; Lubman, D.M., *Rev. Sci. Instr.* **63** (1992) 139–148.
12. Cowen, K.A.; Coe, J.V., *Rev. Sci. Instr.* **61** (1990) 2601–2604.
13. Knorr, F.J.; Chatfield, D.A., in *Proceedings of the 33rd Annual Conference on Mass Spectrometry and Allied Topics*, San Diego (1993) pp. 719–720.
14. Van Breemen, R.B.; Snow, M.; Cotter, R.J., *Int. J. Mass Spectrom. Ion and Phys.* **49** (1983) 35–50.
15. Cotter, R.J. *Biomed. Environ. Mass Spectrom.* **18** (1989) 513–532.
16. Tabet, J.-C.; Cotter, R.J., *Int. J. Mass Spectrom. Ion Processes* **54** (1983) 151.
17. Olthoff, J.K.; Honovich, J.P.; Cotter, R.J., *Anal. Chem.* **59** (1987) 999–1002.
18. Spengler, B,; Cotter, R.J., *Anal. Chem.* **62** (1990) 793–796.
19. Pan, Y.; Cotter, R.J., *Org. Mass Spectrom.* **27** (1992) 3–8.
20. Cotter, R.J., *Anal. Chem.* **64** (1992) 1027A–1039A.
21. Dworshak, R.G.; Ens, W.; Spicer, V.; Standing, K.G., presented at the *43rd ASMS Conference on Mass Spectrometry and Allied Topics*, Atlanta (1995).
22. Dawson, J.H.J.; Guilhaus, M., *Rapid Commun. Mass Spectrom.* **3** (1989) 155.
23. Dodenov, A.F.; Chernushevich, I.V.; Laiko, V.V., *12th Int. Mass Spectrometry Conference, Book of Abstracts*, August 26–31, 1991, p. 153.
24. Coles, J.; Guilhaus, M., *Trends Anal. Chem.* **12** (1993) 203–213.
25. Dodenov, A.F.; Chernushevich, I.V.; Laiko, V.V., in *Time-of-Flight Mass Spectrometry* (Chapter 7), Cotter, R.J. (Ed.), ACS Symposium Series 549, Washington, DC (1994) pp. 108–123.
26. Mirgorodskaya, O.A.; Shevchenko, A.A.; Chernushevich, I.V.; Dodenov, A.F.; Miroshnikov, A.I., *Anal. Chem.* **66** (1994) 99–107.
27. Verentchikov, A.N.; Ens, W.; Standing, K.G., *Anal. Chem.* **66** (1994) 126–133.
28. Myers, D.P.; Li, G.; Yang, P.; Hieftje, G.M., *J. Am. Chem. Soc.* **5** (1994) 1008–1016.
29. Myers, D.P.; Li, G.; Mahoney, P.P.; Hieftje, G.M., *J. Am. Chem. Soc.* **6** (1995) 400–410.
30. Grix, R.; Gruener, U.; Li, G.; Stroh, H.; Wollnik, H., *Int. J. Mass Spectrom. Ion Processes* **93** (1989) 323–330.
31. Boyle, J.G.; Whitehouse, C.M.; Fenn, J.B., *Rapid Commun. Mass Spectrom.* **5** (1991) 400–405.
32. Michael, S.M.; Chien, B.M.; Lubman, D.M., *Rev. Sci. Instr.* **63** (1992) 4277–4284.
33. Chien, B.M.; Michael, S.M.; Lubman, D.M., *Rapid Commun. Mass Spectrom.* **7** (1993) 837–843.
34. Chien, B.M.; Lubman, D.M., *Anal. Chem.* **66** (1994) 1630–1636.
35. Lee, H.; Lubman, D.M., *Anal. Chem.* **67** (1995) 1400–1408.
36. Kaiser, R.E.; Cooks, R.G.; Syka, J.E.P.; Stafford, G.C., *Rapid Commun. Mass Spectrom.* **4** (1990) 30.
37. Doroshenko, V.M.; Cotter, R.J., *Anal. Chem.* **34** (1995) 2180–2187.
38. Louris, J.N.; Cooks, R.G.; Syka, J.E.P.; Kelley, P.E.; Stafford, G.C.; Todd, J.F.J., *Anal. Chem.* **59** (1987) 1677.

39. Beavis, R.C.; Chait, B.T., *Chem. Phys. Lett.* **5** (1991) 479–484.
40. Huth-Fehre, T.; Becker, C.H., *Rapid Commun. Mass Spectrom.* **5** (1991) 378.
41. Whittal, R.M.; Li, L., *Anal. Chem.* **67** (1995) 1950–1954.
42. Brown, R.S.; Lennon, J.J., *Anal. Chem.* **67** (1995) 1998–2003.
43. Zhou, J.; Ens, W.; Standing, K.G.; Verentchikov, A., *Rapid Commun. Mass Spectrom.* **6** (1992) 671–678.
44. Vestal, M.L.; Juhasz, P.; Martin, S.A., *Rapid Commun. Mass Spectrom.* **9** (1995) 1044–1050.
45. Brown, R.S.; Lennon, J.J., *Anal. Chem.* **67** (1995) 3990–3999.

8

Product Ion Spectra from Reflectron Instruments

The reflectron (or ion mirror) introduced in 1974 by Mamyrin[1] was intended to improve mass resolution by compensating for the kinetic energy distribution of ions exiting the ion source. Reflectrons are now common on most commercial MALDI time-of-flight instruments, which generally offer the option of recording mass spectra in the linear or reflectron mode (Figure 8.1). However, most of the MALDI–TOF mass spectra reported in the literature, particularly those in which molecular weight strategies using enzymatic digestion or ladder sequencing are involved, have been obtained in the linear mode. The reason is that considerable fragmentation occurs after ions enter the flight tube. In the linear mode, these metastable fragments are recorded at the same flight times as their precursors, so that sensitivity for molecular weight measurements is higher than in the reflectron mode where they are dispersed, generally unfocused, along the time axis.

At the same time, the time dispersion of metastable products in reflectron instruments is now being effectively exploited to obtain structural information, particularly amino acid sequences for peptides. This may indeed prove to be a far more important role for reflectrons than improvements in mass resolution. The correlated reflex spectra introduced in 1986 by LeBeyec et al.[2] provided a method for recording product ions for techniques (such as plasma desorption) that record discrete product ions that can be correlated with their neutral fragments recorded in the linear detector. The post-source decay approach, described by Spengler and Kaufmann[3], uses ion gating to enable recording of product-ion mass spectra from techniques (in particular, MALDI) which produce an analog ion signal in each time-of-flight cycle. In either case, it is important to note that product ions of different masses are not simultaneously focused by the reflectron, so that provisions have been made on most commercial MALDI–TOF instruments for stepping the reflectron voltage to bring different regions of the product-ion mass spectrum into focus. Other approaches (which are also discussed in this chapter) attempt to focus the entire product-ion mass spectrum simultaneously by carrying out fragmentation in the

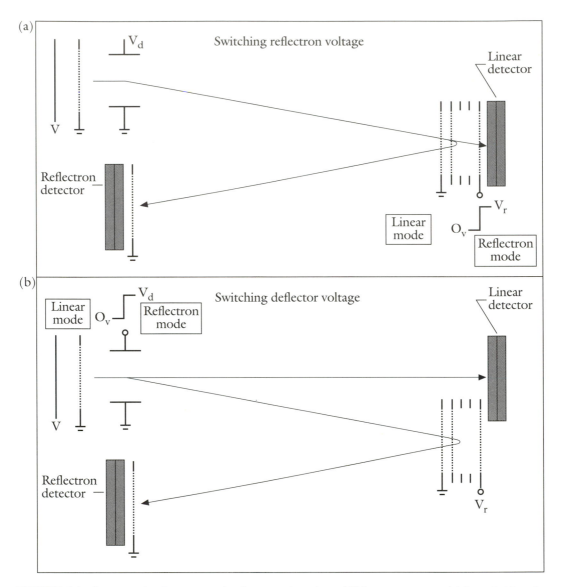

FIGURE 8.1 Linear and reflectron modes for recording time-of-flight mass spectra, (a) by switching the reflectron voltage, and (b) by deflecting the ion beam.

reflectron turn-around region, or through the use of reflectrons with nonlinear electrical fields.

Fragmentation in the Time-of-Flight Mass Spectrometer

When fragmentation occurs promptly (i.e., in the source and within a time frame that is indistinguishable from the time of ionization), the resultant fragment ions are observed at flight times proportional to the square root of their masses. In a linear instrument, flight times are:

$$t = \left(\frac{m}{2eV} \right)^{1/2} D \qquad [1]$$

and in an instrument equipped with a single-stage reflectron they are:

$$t = \left(\frac{m}{2eV} \right)^{1/2} \left[L_1 + L_2 + 4d \right] \qquad [2]$$

While prompt fragmentation is common for small organic ions formed by electron impact, the molecular ions of peptides and proteins formed by desorption techniques fragment over a longer period of time. As described in Chapter 2, metastable fragmentation in the source (as the ions are extracted and accelerated) leads to an incoherent ion signal, tailing of the peaks in the mass spectrum, and a general loss in resolution. These unwanted effects are minimized by using high extraction voltages and a short extraction region, both of which limit the time that ions spend in the accelerating field. This in turn ensures that most of the fragmentation will occur in the drift region, where it is generally referred to as post-source decay. In a reflectron instrument, these product ions are recorded at times that are different from their precursor ions. In a single-stage reflectron instrument, these ions are observed on a linear (rather than a square root) mass scale.

Mass Dispersion of Product Ions in a Single-Stage Reflectron.   If metastable fragmentation occurs in the first field-free drift region (L_1) of a single-stage reflectron instrument:

$$m_a^+ \rightarrow m_b^+ + n \qquad [3]$$

the precursor ions (m_a) will be observed at times:

$$t_a = \left(\frac{m_a}{2eV} \right)^{1/2} \left[L_1 + L_2 + 4d \right] \qquad [4]$$

Product ions (m_b) have the same velocity as their precursor ions in the first drift region (L_1) where fragmentation occurs. Because the reflectron returns ions to the second drift region (L_2) with the same kinetic energy, product ions will also have this same velocity in the second drift region. Thus, the time that product ions spend in the linear regions of the mass spectrometer will be the same as their precursors:

$$t_{b, linear} = \left(\frac{m_a}{2eV} \right)^{1/2} \left[L_1 + L_2 \right] \qquad [5]$$

However, these ions enter (and exit) the reflectron with kinetic energies equal to $(m_b/m_a)eV$, and penetrate the reflectron to a depth of $(m_b/m_a)d$ (where d is the average penetration depth of their precursors), so that the time that product ions spend in the reflectron is given by:

$$t_{b,reflectron} = \left(\frac{m_b}{2\dfrac{m_b}{m_a}eV} \right)^{1/2} 4\frac{m_b}{m_a}d = \left(\frac{m_a}{2eV} \right)^{1/2} 4\frac{m_b}{m_a}d \qquad [6]$$

Thus, the total flight time of a product ion is given by:

$$t_b = \left(\frac{m_a}{2eV} \right)^{1/2} \left[L_1 + L_2 + 4\frac{m_b}{m_a}d \right] \qquad [7]$$

In Chapter 3 we noted that optimal energy focusing for ions formed in the source occurs when the dimensions of the reflectron are such that these ions have an average penetration depth d, where $L_1 + L_2 = 4d$. In this case, precursor ions spend the same time in the linear and reflectron regions. Because product ions have the same velocity as their precursors in the linear region, mass dispersion occurs only during their time in the reflectron, i.e., during one-half of the total flight time of their precursors. Thus, when $L_1 + L_2 = 4d$, the mass scale for product ions can be more conveniently described relative to the flight time of their precursors:

$$t_b = \frac{t_a}{2} \left[\frac{m_b}{m_a} + 1 \right] \qquad [8]$$

This equation describes a linear mass scale that extends from the flight time of the precursor ion to one-half the flight time of the precursor ion, as shown in curve A in Figure 8.2.[4]

Mass Dispersion of Product Ions in a Dual-Stage Reflectron.   The flight times for product ions in an instrument incorporating a dual-stage reflectron depends upon the relative dimensions and electrical field strengths of the two decelerating regions. They are most easily described by considering curve B in Figure 8.2, generated by Standing et al.[4] for a common reflectron design in which precursor ions are decelerated by two-thirds of their kinetic energy in the first retarding region. In this case, product ions having masses below 67% of their precursor-ion masses will not penetrate the second retarding region. Because the first retarding region is short, reflection of these ions is virtually instantaneous, and all of these ions will have nearly identical flight times (the flat portion of curve B). In addition, we noted in Chapter 3 that reflectrons of this particular design focus ions when the average length of the two linear regions is about 8 times the reflectron depth. Because the average velocity of ions during deceleration–acceleration is half that in the drift regions, mass dispersion

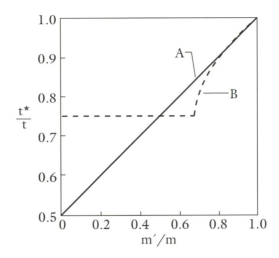

FIGURE 8.2 Calculated flight times for product ions in a single-stage (curve A) and a dual-stage (curve B) reflectron. (Reprinted with permission from reference 4).

occurs for only one-fourth of the flight time of the precursor ion. Thus, the flat line in curve B occurs when $t_b/t_a = 0.75$.

Product ions with masses greater than 67% of their precursor ion's mass are dispersed in time, but not linearly, since the time spent in the first region is not negligible and different for product ions of different mass. As shown in Figure 8.2, mass dispersion of product ions in a dual-stage reflectron is linear only for product-ion masses that are 90 to 100% of the precursor-ion mass, and are dispersed over the last 5% of the precursor-ion flight time. It is for this reason that methods have been developed for obtaining product-ion mass spectra from dual-stage reflectron instruments which record up to 16 mass spectral segments by successively lowering the reflectron voltage by 10%. These are described below.

Coincidence Techniques

In 1986, LeBeyec and co-workers reported a method for recording the product ions formed in a plasma desorption mass spectrometer known as the *correlated reflex technique*.[2] The instrument with which they developed this method is shown in Figure 8.3. It employed a coaxial reflectron and two dual-multichannelplate (MCP) detectors: one for detecting the neutral fragmentation products behind the reflectron and a second detector for recording reflected product ions. With the reflectron voltage on, a neutral fragment, formed by metastable fragmentation in the field-free region, passes through the reflectron, arriving at the first detector at the same time as the precursor ion (or its products) would arrive if the reflectron was turned off. At the same time, the fragment ion associated with this fragmentation event is reflected and arrives at the second detector at a time (since this is a dual-stage design) described by curve B in Figure 8.2. In this method, one first records a linear mass spectrum to determine the flight times of precursor ions. Then, with the reflectron turned on,

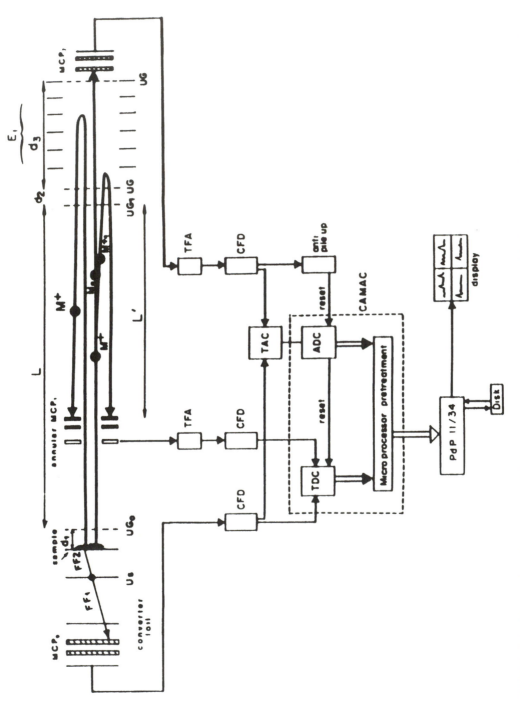

FIGURE 8.3   The coaxial reflectron time-of-flight mass spectrometer of Della-Negra and LeBeyec used for the correlated reflex technique. (Reprinted with permission from reference 2).

product ions are recorded in the second detector only when neutrals arrive at the first detector within a preset time interval, or *mass window*, corresponding to the precursor-ion mass. Thus, all of the product ions recorded are correlated with a particular precursor ion. This method works, of course, only for instruments which record single-ion events. As will be described below, methods which produce large numbers of ions in each ionization event require some form of mass gating.

Figure 8.4a shows a correlated reflex spectrum for the (MH)⁻ ion of a glycosidic terpine (m/z = 461.5), produced by removing the voltage on the second grid so that the reflectron acts a a single-stage reflectron.[5] While the product-ion mass scale is linear in this spectrum, ions of lower mass are increasingly out of focus, so that the peaks corresponding to m/z 100 and 130 are not resolved. In Figure 8.4b, the voltage on the reflectron (now used as a dual-stage reflectron) is lowered to produce a well-resolved mass spectrum of the lower-mass product ions. Thus, this spectrum serves as an early example of the reflectron voltage stepping technique that is now being exploited in post-source decay measurements.

The correlated reflex technique is perhaps best illustrated in an example provided by Standing et al.[6] who utilized this method on a static SIMS time-of-flight mass spectrometer with a single-stage reflectron, or ion mirror. Figure 8.5a shows the linear mass spectrum of the tripeptide GlyGlyPhe with the reflectron turned off, highlighting the ions corresponding to the MH⁺ and MNa⁺ for this molecule for which product mass spectra are to be obtained. Figure 8.5b shows the neutral mass spectrum obtained when the reflectron is turned on, while Figure 8.5c shows the uncorrelated reflected spectrum containing all precursor and product ions. Figures 8.5d and 8.5e then record the product ions when the mass window for neutrals has been chosen to include those neutrals having flight times corresponding to MH⁺ and MNa⁺ ions, respectively.

An additional example of the correlated technique is also provided by Standing and co-workers.[7] Figure 8.6a shows the linear time-of-flight mass spectrum of the molecular-ion region of tyrothricin, recorded in the detector located behind the reflectron. When the reflectron is turned on and the spectrum recorded in the second detector, the MH⁺, (MH+14)⁺, and MNa⁺ ions for the tyrocidin A, tyrocidin B, and tyrocidin C components of this mixture are well resolved (Figure 8.6b). At the same time, the mass spectrum of the neutrals recorded in the first detector (Figure 8.6c) illustrates the fact that many of these molecular ions undergo metastable decay in the flight tube. While the spectrum in Figure 8.6c is poorly resolved, it is possible to center a time window on the peaks corresponding to flight times for the MH⁺ ions of tyrocidin A and tyrocidin B to produce the correlated product-ion mass spectra shown in Figures 8.6d and 8.6e, respectively.

Post-Source Decay for Peptide Sequencing

While an effective approach, the correlated reflex (or coincidence) technique can be utilized only on instruments which record single-ion events (primarily plasma-desorption mass spectrometers), and not on instruments which record (and digitize) analog transients in each time-of-flight cycle. Thus, for instruments using matrix-assisted laser desorption/ionization (MALDI) it becomes necessary to utilize an ion-gating system in the first drift region (L_1) to distinguish product ions arising from different molecular species, or from matrix ions that are overlaid on the same time axis (Figure 8.7). No matter how well-designed, a gating system located in the first

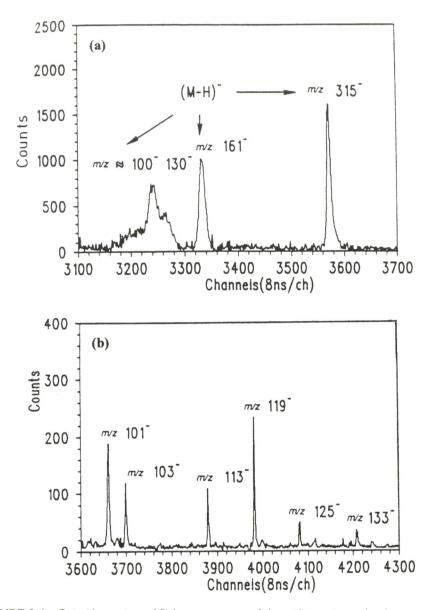

FIGURE 8.4 Coincidence time-of-flight mass spectra of glycosidic terpine molecules correspond-ing to the in-flight decay of the molecular anion (M-H)⁻ at m/z 461.5. (a) Mass spectrum obtained at full reflectron voltage with the reflectron configured as a single-stage reflectron. (b) Mass spectrum obtained at lowered reflectron voltage to focus the lower mass region of the spectrum. In this case, the reflectron is operated in a dual-stage configuration and results in mass resolution of around 3000. (Reprinted with permission from reference 5).

drift region will have limited mass selectivity, since mass dispersion is low at this point and the ions are brought to full focus only at the end of the second linear region (L_2). As will be described in Chapter 11, high performance mass selectivity can be achieved in tandem instruments when gating can be performed at locations in which

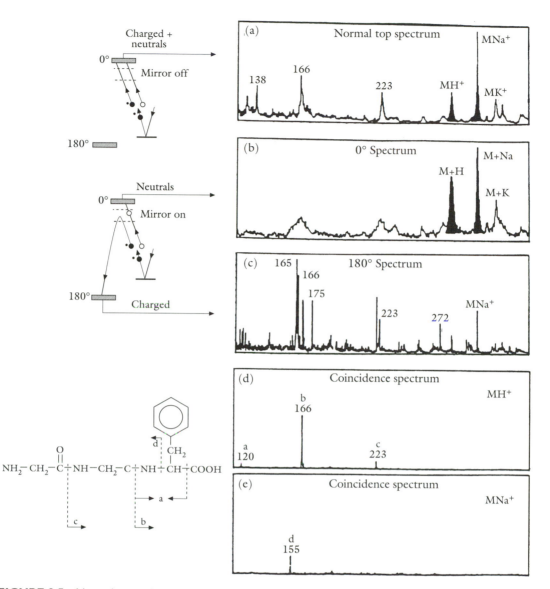

FIGURE 8.5 Normal, neutral, and correlated spectra of GlyGlyPhe obtained on a static SIMS time-of-flight mass spectrometer with a single-stage reflectron. (Reproduced with permission from reference 6).

the ions are brought into focus by the first reflectron. However, the "poor man's tandem" configuration provided by a single reflectron instrument has been very effective for obtaining amino acid sequences for peptides.

Post-Source Decay Spectra Using a Dual-Stage Reflectron.   The discussion on coincidence techniques above suggests that the opportunities for recording product ion spectra on a reflectron instrument and the necessity for obtaining spectra at different reflectron voltages have been appreciated for some time. Nevertheless, the currently popular technique known as post-source decay (PSD) is based upon an

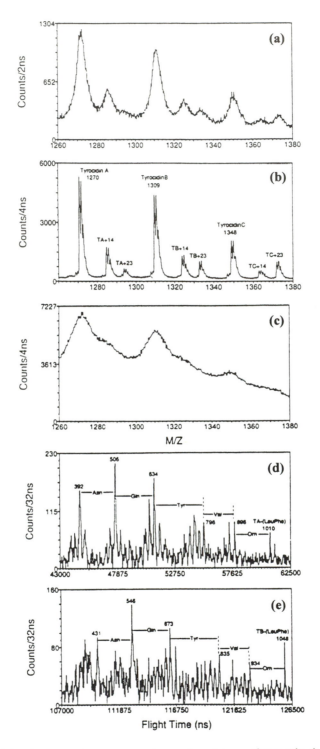

FIGURE 8.6 (a) Molecular-ion region mass spectrum of a mixture of tyrocidin A, B, and C, obtained in the linear mode with the reflectron voltage turned off, (b) molecular-ion region obtained in the reflectron mode, (c) neutral mass spectrum of the molecular-ion region with the reflectron voltage turned on, (d) correlated product-ion mass spectrum of tyrocidin A, and (e) correlated product-ion mass spectrum of tyrocidin B. (Reprinted with permission from reference 7).

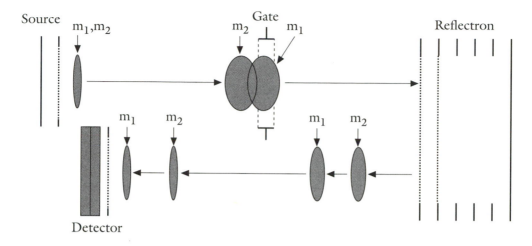

FIGURE 8.7   Diagram of a reflectron instrument showing the location of an ion-gating system.

approach introduced by Spengler and Kaufmann,[3] in which a large number of spectral segments are obtained from a dual-stage reflectron instrument by stepping the reflectron voltages. Because metastable fragmentation can occur both in the source–accelerating region (which is basically unproductive) as well as in first linear region (L_1) of the mass analyzer, the term post-source decay was introduced to describe fragmentation processes occuring after the ions leave the ion source, including those formed by metastable decomposition or by opportunistic collisions with residual gas molecules in the vacuum chamber.

Figure 8.8 is a schematic of the flight times and mass spectra that would be observed for molecular and product ions in a dual-stage reflectron instrument operated a full reflectron voltage, that is, when U_2 is slightly larger than U_{acc} (the accelerating voltage).[8] In this configuration, molecular ions (MH[+]) penetrate the reflectron to the optimal depth and are fully focused. Product ions ranging in mass from 0.63 to 1.0 of the molecular-ion mass enter the second stage of the reflectron and are also focused, but only those from 0.9 to 1.0 of the molecular-ion mass (as described in Figure 8.2) are focused on a linear scale. Product ions with masses below 0.63 of the molecular-ion mass do not penetrate the second retarding region and are not focused. They appear as a single, unresolved peak designated as small PSD fragments. In effect, the useful region is that encompassing ions with masses 0.9 to 1.0 times the molecular-ion mass, or (since it is the energy range rather than mass range that is focused) ions with energies in the range of 0.9 to 1.0 eU_{acc}. Thus, other regions of the mass spectrum become accessible by lowering the reflectron voltage.

Figure 8.9 shows the mass spectral segments for Substance P recorded by successive stepping of the reflectron voltage to 85% of its previous value. In the first spectrum, the accelerating voltage U_{acc} is 10 kV and the reflectron voltage U_2 is 10.8 kV. In subsequent spectra, the reflectron voltage is lowered to 8640 V and 6912 V etc., with the voltage on the first stage reduced proportionately such that $U_1 = 0.57\ U_2$. The fourteen linear segments are then combined in Figure 8.10 to form a composite spectrum of Substance P that reveals most of the major amino acid sequence-specific fragment ions.

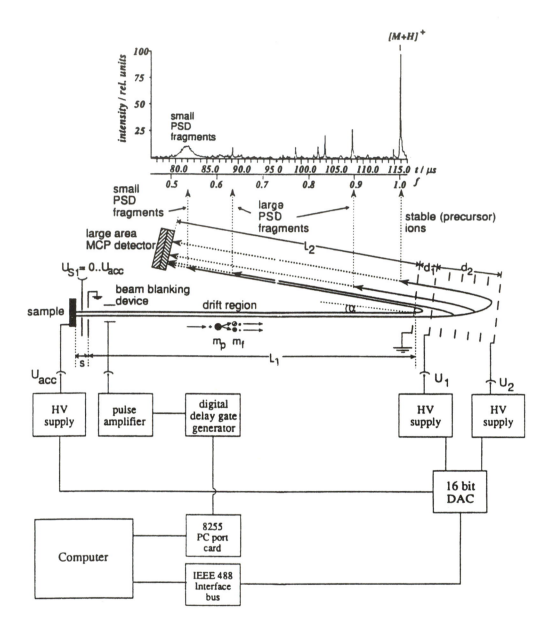

FIGURE 8.8 The post-source decay scheme for a dual-stage reflectron instrument. (Reprinted with permission from reference 8).

Interestingly, no gating system is required in this scheme when product ions are obtained for a single peptide. In the mass spectrum obtained at full reflectron voltage, matrix ions appear in the lower-mass region of the spectrum and not in the spectral segment used to construct the final mass spectrum. As the reflectron voltage is lowered, these ions (which have full kinetic energy due to acceleration) pass through the reflectron and (therefore) do not interfere with the post-source decay spectrum.

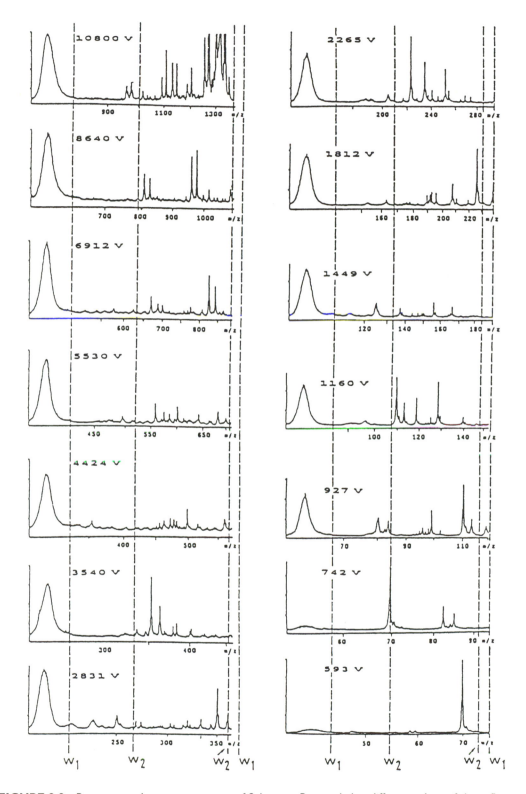

FIGURE 8.9 Post-source decay mass spectra of Substance P recorded at different values of the reflectron voltage. (Reprinted with permission from reference 8).

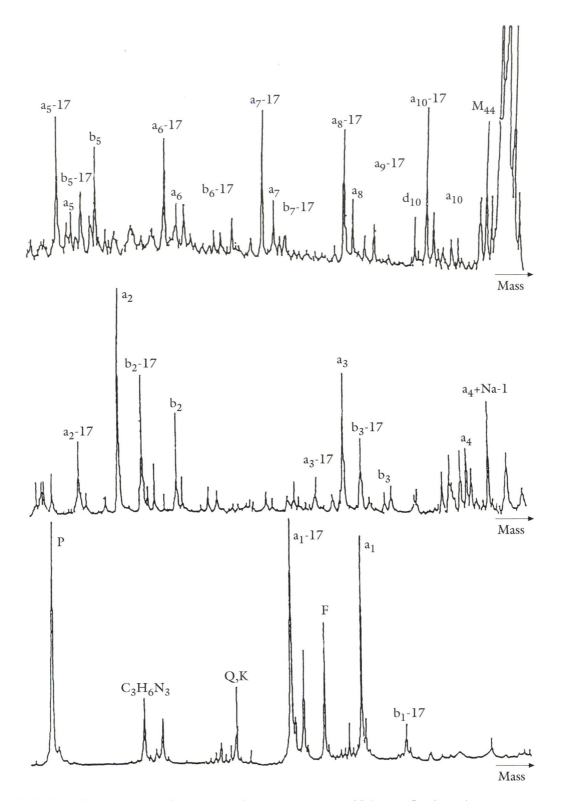

FIGURE 8.10   Reconstructed post-source decay mass spectrum of Substance P utilizing the upper regions of the spectra shown in Figure 8.9. (Reprinted with permission from reference 8).

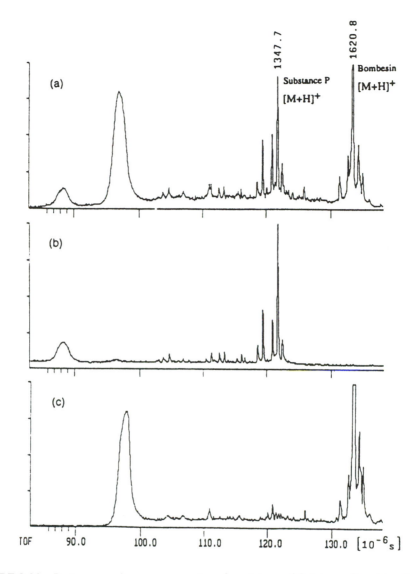

FIGURE 8.11 Post-source decay mass spectra of a mixture of Substance P and bombesin. (a) Mass spectrum without precursor-ion selection at full reflectron voltage. (b) Selection of the MH⁺ ion of Substance P. (c) Selection of the MH⁺ ion of bombesin. (Reprinted with permission from reference 3).

When more than one peptide is present, ion gating is required as shown in Figure 8.11.[3] In these spectra, obtained at full reflectron voltage, both focused and small PSD fragments are observed for each component, but can be separated by the application of an ion gate. Figure 8.12 shows the assembled product-ion mass spectrum of bombesin obtained from this mixture from 12 mass-spectral segments.

While these are impressive results from a relatively simple and inexpensive instrument, the obvious downside to this approach is the need to acquire multiple spectra in which only a small portion of the ion signal from each is utilized to construct the

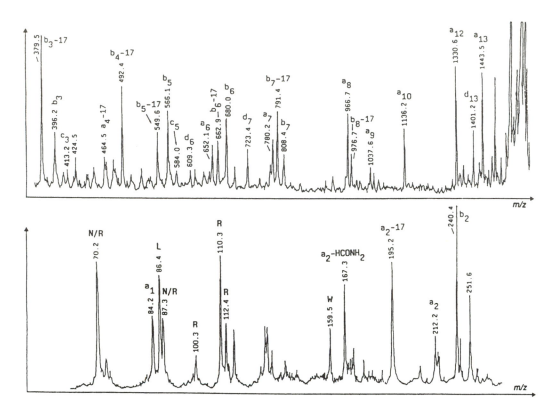

FIGURE 8.12   Reconstructed post-source decay mass spectrum of bombesin obtained from 12 spectral fragments following selection of the MH⁺ ion for bombesin in the mixture shown in Figure 8.11. (Reprinted with permission from reference 3).

final product-ion mass spectrum. Thus, this approach loses much of the multichannel recording advantage of the time-of-flight instrument and limits the overall sensitivity. In addition, because the mass scale for each of the spectral segments have different (linear) slopes, it is generally necessary to acquire calibration spectra for each of the 12 to 14 values of the reflectron voltage. These problems are alleviated somewhat by the use of a single-stage reflectron, which requires fewer reflectron voltage steps to focus all of the ions. Such approaches are discussed below. In addition, there have also been efforts to achieve simultaneous focusing over a broad mass range using reflectrons with nonlinear retarding fields.

Post-Source Decay Spectra Using a Single-Stage Reflectron.   Dual-stage reflectrons provide second-order focusing for molecular ions and prompt fragment ions, but focus only those product ions formed by post-source decay (approximately the upper 30% of the molecular-ion mass range) that enter the second retarding region. In addition, the useful region (in which the mass scale is linear) is limited to the upper 10% of the precursor-ion mass range. While dual-stage reflectrons undoubtably provide better focusing than single-stage reflectrons in this short mass range, single-stage reflectrons provide

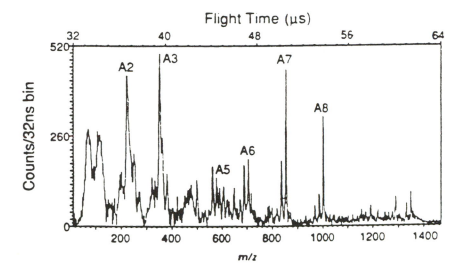

FIGURE 8.13 Product-ion mass spectrum resulting from the post-source decay of the MH⁺ ion of Substance P with the reflectron voltage optimized for the precursor ion. (Reprinted with permission from reference 9)

an opportunity to acquire post-source decay spectra in many fewer steps since all product ions enter the linear retarding region.

An example by Standing et al.[9] illustrates this point. Figure 8.13 shows the post-source decay mass spectrum of the MH⁺ ion of substance P (m/z = 1347) obtained on a SIMS–TOF using coincidence techniques. Compared with the PSD spectrum of substance P in Figure 8.11b, the loss of resolution toward lower mass is more gradual and, in addition, the mass scale is linear throughout. However, a better resolved PSD mass spectrum can be constructed from the five mass spectral segments (obtained at different reflectron voltages) shown in Figure 8.14.

Figure 8.15 shows the post-source decay mass spectrum of thioredoxin obtained by Vestal at al.[10] on a single-stage reflectron time-of-flight mass spectrometer using only two reflectron voltage steps. In principle, because the mass scale for product ions using a single-stage reflectron is linear, the entire product-ion mass range should be observed in reasonable focus if the ions require minimal correction for the kinetic energy spread. In practice, it is difficult to achieve such focusing in one or two segments, since the laser power must generally be increased well above threshold to induce post-source decay, and this results in increases in ion kinetic energy that require correction by the reflectron.

The mass spectra in Figure 8.16 obtained by Martin et al.[11] illustrate how post-source decay mass spectra might be used in the complete characterization of a unknown protein. Figure 8.16a is the linear-mode MALDI–TOF mass spectrum of the intact protein that establishes the approximate molecular weight at 24 kDa. The protein is then digested with trypsin and reanalyzed to produce the *tryptic map* shown in Figure 8.16b. Using the gating system, one of the peptides is mass-selected, the laser power is increased, and the PSD mass spectrum obtained as shown in Figure 8.16c.

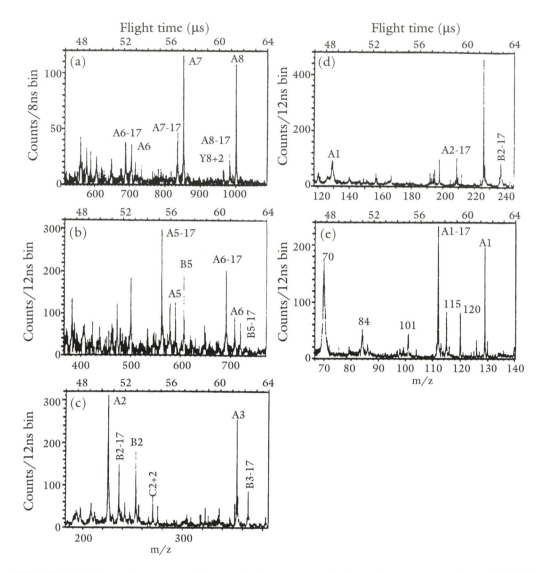

FIGURE 8.14 Focused segments of the product-ion mass spectra from the post-source decay of the MH$^+$ ion of Substance P with the reflectron voltage optimized for (a) the A$_8$ ion at *m/z* 1001, (b) the A$_6$ ion at *m/z* 707, (c) the A$_3$ ion at *m/z* 354, (d) the A$_2$ ion at *m/z* 226 and (e) the A$_1$ ion at *m/z* 129. (Reprinted with permission from reference 9).

Calibrating Post-Source Decay Mass Spectra. Methods for calibrating post-source decay mass spectra and for determining the masses of product ions for a peptide of unknown sequence are relatively straightforward for instruments using a single-stage reflectron. Figure 8.17 is a graphical representation of the mass scales for normal and post-source decay mass spectra of an unknown sample m_a (*m/z* = 2000) and a calibration compound m_c (*m/z* = 1000), which may be present in the sample mixture (if the instrument is equipped with a mass-gating system) or used as

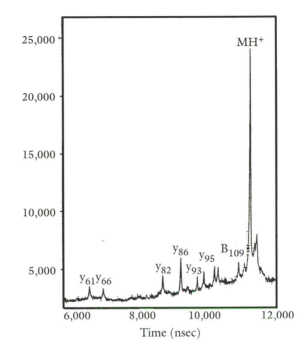

FIGURE 8.15 Post-source decay mass spectrum of *E. coli* thioredoxin obtained from a single-stage reflectron TOF mass spectrometer in two reflectron voltage segments. (Reprinted with permission from reference 10).

an external standard. In either case, both of the precursor ions (m_a and m_c) lie along the line:

$$t = km^{1/2}$$ [9]

that is, the flight time of the unknown m_a is:

$$t_a = \left(\frac{m_a}{2eV}\right)^{1/2}\left[L + 4d\right]$$ [10]

while that of the calibration compound m_c is:

$$t_c = \left(\frac{m_c}{2eV}\right)^{1/2}\left[L + 4d\right]$$ [11]

where $L = L_1 + L_2$, the sum of the lengths of the linear drift regions. The calibration compound forms product ions:

$$m_c^+ \rightarrow m_d^+ + n$$ [12]

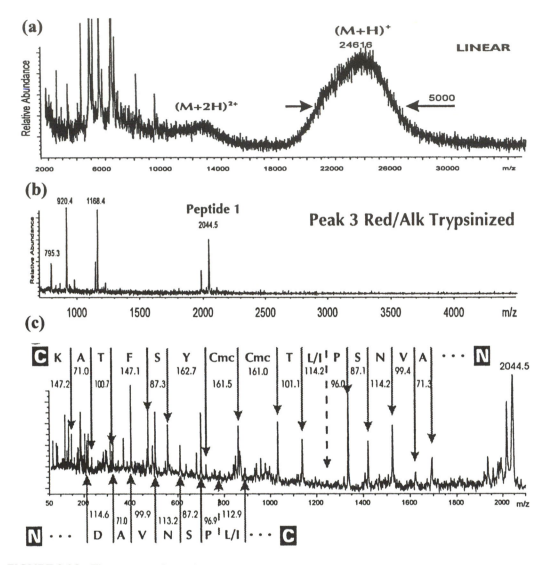

FIGURE 8.16 Three views of an unknown protein obtained by MALDI–TOF mass spectrometry: (a) linear-mode mass spectrum of the intact protein, (b) mass map of the tryptic peptides derived from the unknown protein, and (c) PSD mass spectrum of peptide 1 from the mixture in (b). (Reprinted with permission from reference 11).

whose flight times are linear in mass:

$$t_d = \left(\frac{m_c}{2eV}\right)^{1/2}\left[L + 4\frac{m_d}{m_c}\,d\right]$$

[13]

and extend from the flight time of the precursor (when $m_d = m_c$) to:

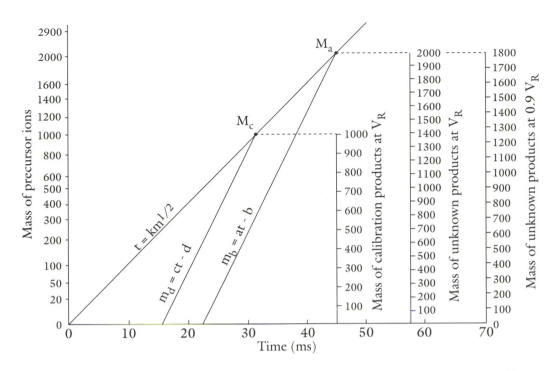

FIGURE 8.17 Schematic representation of the normal and post-source decay mass scales for a calibration compound m_c and a sample m_a to be calibrated. Precursor ions are observed along the line: $t = km^{1/2}$ using the left-hand (square root) scale. Product ions of m_c and m_a are observed along the lines m_d and m_b, respectively, using the linear mass scales for full V_R and reduced $0.9V_R$ reflectron voltage shown on the right. The constants a-d are described in the text.

$$t_d = \left(\frac{m_c}{2eV} \right)^{1/2} L \qquad [14]$$

when $m_d = 0$. Since single-stage instruments are generally designed such that $L = 4d$, this is approximately one half the flight time of m_c. Similarly the product ions from the unknown compound:

$$m_a^+ \rightarrow m_b^+ + n \qquad [15]$$

have flight times corresponding to:

$$t_b = \left(\frac{m_a}{2eV} \right)^{1/2} \left[L + 4\frac{m_b}{m_a} d \right] \qquad [16]$$

that also form a linear mass scale extending from the flight time to approximately one-half the flight time of its precursor ion m_a. The relationship between the calibration and unknown product-ion mass scales at full reflectron voltage V_R is shown graphically in Figure 8.17.

Although single-stage reflectrons are generally designed so that a precursor ion spends equal times in the linear and reflectron regions (i.e., $L = 4d$), the total flight time also includes the time spent in the source region, the time spent to reach the space focus plane, etc. In addition, the voltage applied to the reflectron V_R, tolerances for the resistors used to set the voltages on the reflectron lenses, etc., effect the actual turn-around depth d and (consequently) the focal length in the linear regions. Assuming that all of these effects are included in the above equations, then the ratio $L/4d = k'$ can be used to express the ratio of the time that product ions spend in all regions in which their velocities are the same as that of their precursors to the time during which they are dispersed according to mass. The calibration equation can then be obtained by dividing equation 12 by equation 11:

$$\frac{t_d}{t_c} = \frac{k' + \dfrac{m_d}{m_c}}{k' + 1} \qquad [17]$$

or more simply as:

$$\frac{m_d}{m_c} = (k' + 1)\frac{t_d}{t_c} - k' \qquad [18]$$

where $k'+1$ and k' are the slope and intercept, respectively, of the calibration equation determined by the set of time t_d and mass m_d data points in the calibration PSD mass spectrum. Thus, the primary result of the calibration data is the determination of a single parameter k', that can be used to calibrate the masses of all product ions, since the fraction of time spent in the linear and reflectron regions is the same for all precursor ions. Provided that no instrumental parameters are changed (such as, for example, the reflectron voltage), the mass of product ions m_b from the unknown m_a can be determined from the equation:

$$\frac{m_b}{m_a} = (k' + 1)\frac{t_b}{t_a} - k' \qquad [19]$$

Thus, when equations 18 and 19 are used, which express the ratio of product to precursor mass as a function of the ratio of their flight times, the data for both calibration and unkown PSD mass spectra will lie on the same line. If (however) we wish to express the relationship between actual mass and flight time, the product-ion masses from the calibration and unknown PSD mass spectra will be given by:

$$m_d = ct_d - d \qquad [20]$$

where $c = (k'+1)/m_c t_c$ and $d = k'/m_c$, and:

$$m_b = at_b - b \qquad [21]$$

where $a = (k'+1)/m_a t_a$ and $b = k'/m_a$, respectively. These are plotted in Figure 8.17, and describe the relationship between the PSD and normal mass spectra.

Mass Calibration at Reduced Reflectron Voltages.   If the reflectron voltage is then lowered by 10% to 0.9 V_R, molecular ions m_a will pass through the reflectron and (as shown in Figure 8.17) product-ion masses in the range of 0 to 0.9 m_a will be reflected. In principle, it should be possible to assign masses to the peaks in this PSD spectrum utilizing the same calibration obtained at full reflectron voltage. In this case, the mass of each product ion would simply be 0.9 m_b, where m_b is the mass determined from either equation 19 or 21. However, there are a number of difficulties with this approach, the first of which is that it demands that each new voltage value be set accurately and reproducibly. This may require some settling time for the resistors connecting the retarding lenses which now operate at slightly different temperatures. In addition, some changes in actual resistance may in fact alter the effective distance d at each voltage value, thus affecting the value of k' used in the calibration equations. For these reasons, it is common to utilize an internal calibration standard, and to obtain a new calibration spectrum for each value of the reflectron voltage. The result, of course, is that obtaining PSD mass spectra for each component of an unknown mixture can be a time-consuming process. Thus, there has been considerable interest in approaches which record the entire product-ion mass spectrum at full reflectron voltage. These approaches (as described below) include (1) photodissociation in the turn-around region, and (2) nonlinear reflectrons.

Mass calibration for instruments using dual-stage reflectrons is, of course, similar if one utilizes only those portions of the post-source decay spectra that are linear.

Photodissociation at the Turn-Around Region.   Duncan and co-workers[12] have described an instrument (shown in Figure 8.18) in which ions are photodissociated inside the reflectron, just as these ions reach zero velocity and are turning around to be reaccelerated out of the reflectron. This scheme is of considerable advantage when the method for dissociation involves a short (5 ns) pulsed laser, since the low ion velocity and the shape of the ion trajectories at this point insures the maximum temporal and spatial overlap with the laser beam. In addition, since product ions are formed with nearly zero kinetic energies (i.e., there is an uncorrected initial kinetic energy spread at the turn-around region), the dual-stage reflectron used in their instrument in fact acts as a dual-stage extraction region with the turn-around point acting as a source, and the flight times become proportional to the square root of mass:

$$t_b = am_a^{1/2} + bm_b^{1/2} \qquad [22]$$

where the first term is the time for the precursor ion m_a to travel from the source to the turn-around region (and is used to determine the time for firing the photo-dissociating laser), and the second term is the time it takes for the product ion to reach the detector.

This instrument developed by Duncan et al.[12] was used with a metal-ion cluster source to study the photodissociation of Sn_6^+ clusters, while a similar instrument was reported by Ledingham and co-workers[13] for the photodissociation of organic ions

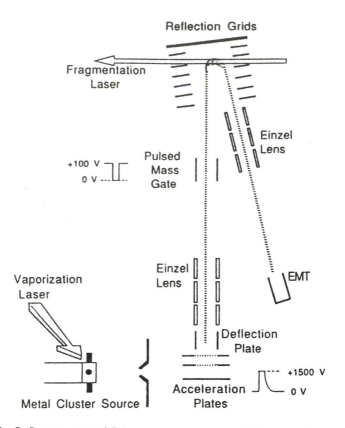

FIGURE 8.18   Reflectron time-of-flight mass spectrometer of Duncan and co-workers.[12] Metal ion clusters are formed in a cluster-ion source and photodissociated in the turn-around region of the reflectron. (Reprinted with permission from reference 12).

using the 266- and 355-nm outputs from a Nd:YAG laser. Figure 8.19 shows their product-ion mass spectrum for the $(M-OH)^+$ ion of the MALDI matrix compound 2,5 dihydroxybenzoic acid (DHB), desorbed and ionized from a solid sample using a pulsed nitrogen (337 nm) laser.

In principle, this approach avoids all of the problems associated with the different penetration depths of ions fragmented prior to entering the reflectron. Because all precursor ions enter and all product ions exit the reflectron with kinetic energies of approximately eV (where V is the initial accelerating voltage), they are all equally focused at the same reflectron voltage. At the same time, mass resolution is limited by the fact that the product ions are, in effect, being focused by a linear mass analyzer. The turn-around region acts as a source, in which the initial kinetic energies of the precursor ions have been converted entirely to a spatial distribution (all the ions have zero velocity). The detector is then located at the space-focus plane. In addition, this approach requires that fragmentation be prompt, since the reflectron itself provides a very long extraction region. Thus, this configuration would be less suitable for the fragmentation of peptides (which generally does not occur promptly) or for collision-induced dissociation (which would not be effective at near zero velocities).

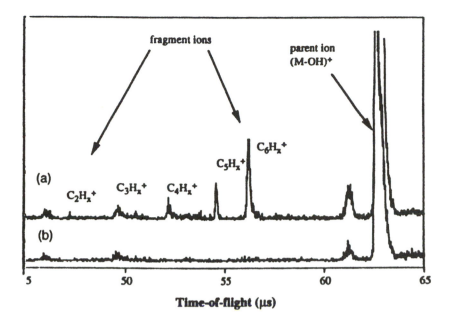

FIGURE 8.19   (a) Photodissociation mass spectra of the (M-OH)$^+$ ion of DHB with a 266-nm Nd:YAG laser focused on the turn-around region of the reflectron, and (b) mass spectrum with the photodissociation laser turned off. (Reprinted with permission from reference 13).

Nonlinear Reflectrons.   Therefore, while dissociation in the turn-around region provides simultaneous focusing, the post-source decay approach has proven to be far more effective for the structural analysis of biological molecules, but requires the wasteful and time-consuming process of recording sample and calibration mass spectra at a number of different reflectron voltages. One attempt to alleviate this problem is the curved-field reflectron designed by Cotter and Cornish.[14] In this reflectron, the voltages placed on the reflectron lens plates follow a portion of the equation for a circle (Figure 8.20), resulting in a nonlinear retarding field that provides focusing of product ions over a broader range of mass (and energy).

     This reflectron has been utilized in an instrument (Figure 8.21) in which a coaxial dual-channelplate detector is placed at the exit of the reflectron (to maximize the distance L_1 in which post-source decay occurs by making $L_2 = 0$), a gate has been provided for selection of the precursor ion, and a pulsed valve is included for collision-induced dissociation. Figure 8.22 shows the curved-field mass spectra of the peptide dermenkephalin. In Figure 8.22a, the ion gate is turned off and the mass spectrum includes peaks originating from both the peptide and matrix ions. In Figure 8.22b, the ion gate is used to select the protonated molecular ions of the peptide. In this spectrum, the baseline signal arises from fragmentation occuring in the reflectron itself. In Figure 8.23, this background is subtracted revealing a complete set of amino acid sequence-specific ions, which are in focus throughout the mass range.[14]

     Mass calibration is relatively straightforward even though it does not follow a simple linear or square-root mass law. Figure 8.24 shows the product-ion mass spectrum obtained on the curved-field reflectron TOF for the synthetic peptide

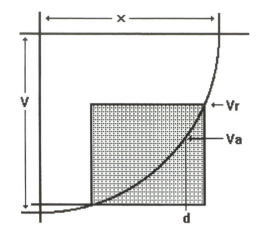

FIGURE 8.20 Schematic representation of the method used to determine the voltages on the retarding ring electrodes of a curved-field reflectron. The voltages placed on the reflectron lenses are determined from the equation of a circle: $R^2 = V^2 + x^2$, where V is a voltage, x is distance, and R is a constant. The portion of the circle utilized in the reflectron is shown in the shaded area, where V_R is the voltage on the back of the reflectron and V_a is the accelerating voltage that locates the turn-around depth.

$(Pro)_{14}Arg$, which provides a series of y-ions for mass calibration. The time and mass data from this spectrum is then fitted to the equation of an ellipse to provide a mass calibration curve (such as shown in Figure 8.25). This curve plots the ratio of product–precursor masses as a function of the ratio of product–precursor flight times, and (since the fraction of times that precursor ions spend in the linear and reflectron fields is constant) all PSD mass spectra lie on this same curve. Therefore, because the reflectron voltage does not have to be changed, PSD mass spectra are all obtained at a single reflectron voltage using a single calibration. More remarkable is the fact that, if the reflectron and accelerating voltages are derived from the same supply, the PSD calibration does not have to be redetermined for each mass spectrum, each sample, or each time that the instrument is powered up. Because the ratio V_R/V_a remains constant, an initial calibration describes an instrumental parameter that remains unchanged for every PSD mass spectrum. Thus, for example, the mass for a molecular MH$^+$ ion of an unknown would first be calibrated by the normal practice using two known masses. However, its PSD mass spectrum is calibrated by a simple change in the mass scale, since the PSD mass scale is directly derived from the square-root scale. Figure 8.26 shows the calibrated mass spectrum of *angiotensin II* obtained in this manner.[15]

Comparison of Reflectrons Used for Post-Source Decay.   Figure 8.27 shows the voltage profiles for time-of-flight mass spectrometers equipped with several different kinds of reflectrons. In all cases, ions begin their flight from an ion source held at the accelerating voltage V_a, and the focal point for the reflectron analyzers is the space-focus plane at a distance equal to 2X the width of the source. L_1 and L_2

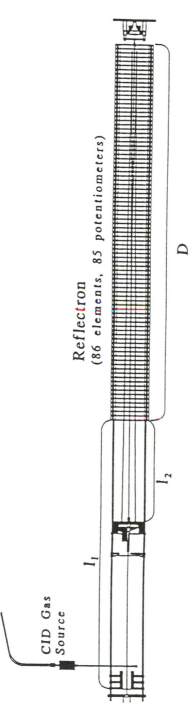

FIGURE 8.21   Coaxial, curved-field reflectron TOF with gate and pulsed valve shown. (Reprinted with permission from reference 15).

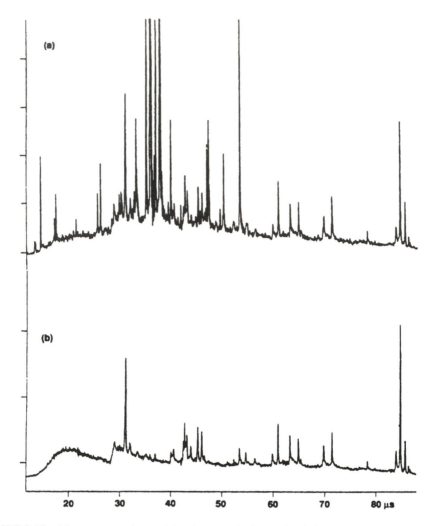

FIGURE 8.22 Mass spectra obtained for the peptide dermenkephalin using a curved-field reflectron. (a) Ion gate OFF. (b) Ion gate ON. (Reprinted with permission from reference 14).

are used to designate the linear regions traversed by the ions before and after they enter the reflectron, and are assumed to be equal in these examples. The voltage at the rear of the reflectron is V_R and is somewhat larger than V_a. Precursor ions are reflected when they reach the point d in the reflectron where the voltage is V_a, while product ions are reflected at shorter depths.

We have noted (Chapter 3) that when used to focus precursor ions, the dual-stage reflectron provides second-order focusing. A similar advantage exists for product ions formed by post-source decay; however, the useful focusing region is limited to about the top 10% of the spectrum, necessitating the acquisition of many more spectral segments than for the single-stage reflectron. We have also noted that dual-stage reflectrons can be designed that are considerably smaller in size than single-stage reflectrons, since their focal lengths $(L_1 + L_2)$ are much longer. For an instrument

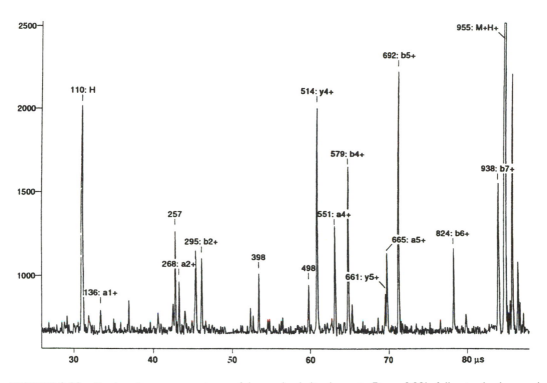

FIGURE 8.23   Product-ion mass spectrum of dermenkephalin shown in Figure 8.22b, following background subtraction. (Reprinted with permission from reference 14).

of fixed size, this provides a distinct advantage for the post-source decay technique, since the length (L_1) over which metastable decay occurs is considerably longer.

The quadratic-field reflectron provides (in theory) focusing that is independent of kinetic energy, but the low ion-transmission characteristics resulting from electrical field gradients in the off-axis direction have made this reflectron impractical. In addition, because the focal length of this reflectron is zero (i.e., there are no linear regions), it is also impractical for carrying out post-source decay measurements. The curved-field reflectron described above utilizes linear regions which are shorter than those on single- or dual-stage reflectron analyzers. Therefore, in order to extend the first linear region L_1 and improve post-source decay efficiency, a coaxial arrangment in which the detector is located at the exit of the reflectron (i.e., $L_2 = 0$) is the most practical. Because of its similarity to the quadratic reflectron curvature, the curved-field reflectron provides somewhat better focusing for precursor ions than single-stage reflectrons, and (as described above) permits recording of post-source decay mass spectra without scanning the reflectron voltage V_R.

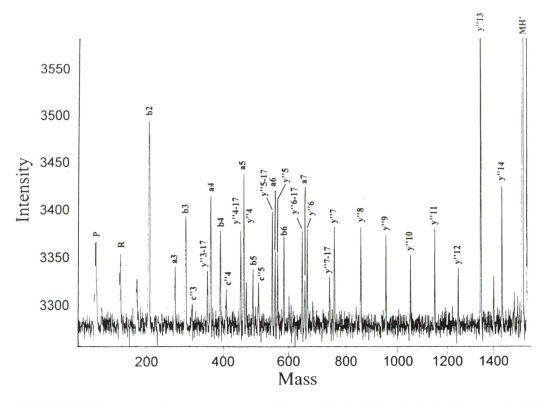

FIGURE 8.24 Post-source decay mass spectrum of the peptide (Pro)$_{14}$Arg obtained using a curved-field reflectron and used for mass calibration. (Reprinted with permission from reference 15)

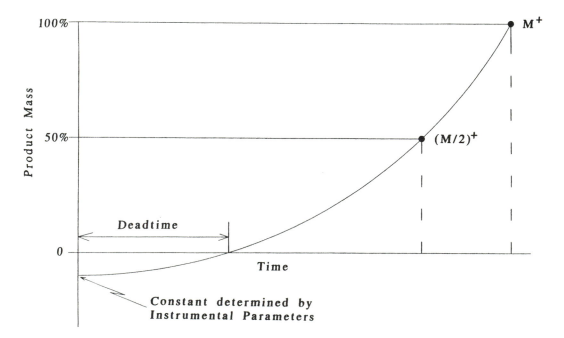

FIGURE 8.25 General form of the calibration curve used for calibrating post-source decay mass spectra on a curved-field reflectron TOF. (Reprinted with permission from reference 15).

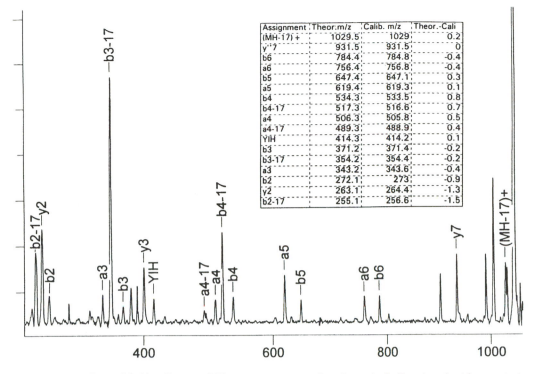

Assignment	Theor. m/z	Calib. m/z	Theor.-Cali
(MH-17) +	1029.5	1029	0.2
y"7	931.5	931.5	0
b6	784.4	784.8	-0.4
a6	756.4	756.8	-0.4
b5	647.4	647.1	0.3
a5	619.4	619.3	0.1
b4	534.3	533.5	0.8
b4-17	517.3	516.6	0.7
a4	506.3	505.8	0.5
a4-17	489.3	488.9	0.4
YIH	414.3	414.2	0.1
b3	371.2	371.4	-0.2
b3-17	354.2	354.4	-0.2
a3	343.2	343.6	-0.4
b2	272.1	273	-0.9
y2	263.1	264.4	-1.3
b2-17	255.1	256.6	-1.5

FIGURE 8.26   Curved-field reflectron PSD mass spectrum of angiotensin *II*. (Reprinted with permission from reference 15).

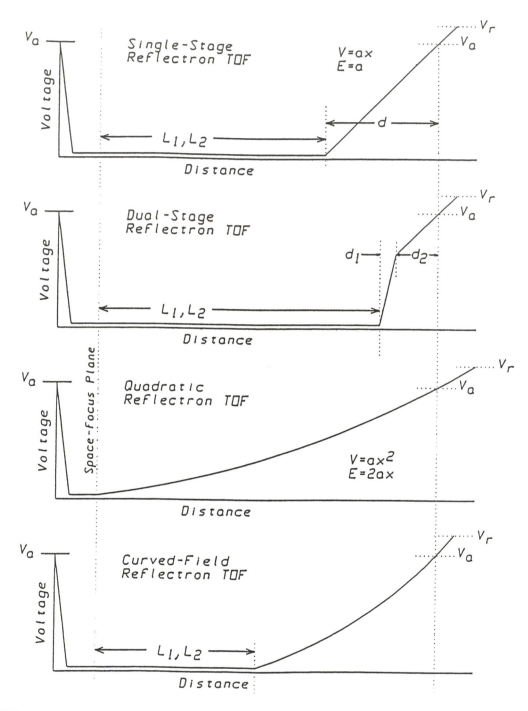

FIGURE 8.27 Comparison of the voltages in time-of-flight mass spectrometers using (a) a single-stage reflectron, (b) a dual-stage reflectron, (c) the curved-field reflectron, and (d) a quadratic reflectron.

References

1. Mamyrin, B.A.; Karatajev, V.J.; Shmikk, D.V. and Zagulin, V.A., *Soc. Phys. JETP* **37** (1973) 45–48.
2. Della-Negra, S. and LeBeyec, Y., in *Ion Formation from Organic Solids IFOS III*, Benninghoven, A., Ed.; (1986) pp 42–45.
3. Kaufmann, R.; Spengler, B. and Lutzenkirchen, F., *Rapid Commun. Mass Spectrom.* **7** (1993) 902–910.
4. Tang, X.; Beavis, R.; Ens, W.; LaFortune, F.; Schueler, B. and Standing, K.G., *Int. J. Mass Spectrom. Ion Processes* **85** (1988) 43–67.
5. Brunelle, A.; Della-Negra, S.; Depauw, J.; Joret, H.; LeBeyec, Y., *Rapid Commun. Mass Spectrom.* **5** (1991) 43.
6. Standing, K.G.; Beavis, R.; Bolbach, G.; Ens, W.; LaFortune, F.; Main, D.; Schueler, B.; Tang, X.; Westmore, J.B., *Anal. Instrumen.* **16** (1987) 173.
7. Standing, K.G.; Ens, W.; Mao, Y.; Lafortune, F.; Mayer, F.; Poppe, N.; Schueler, B.; Tang, X. and Westmore, J.B., *J. de Physique* **C2** (1989) 163.
8. Kaufmann, R.; Kirsch, D.; Spengler, B., *Int. J. Mass Spectrom. Ion Processes* **131** (1994) 355–385.
9. Tang, X.; Ens, W.; Mayer, F.; Standing, K.G. and Westmore, J.B., *Rapid Commun. Mass Spectrom.* **3** (1989) 443–448.
10. Vestal, M.L.; Blakeley, C.R.; Verentchikov, A., in *Proceedings of the 42nd ASMS Conference on Mass Spectrometry and Allied Topics*, Chicago (1994) p. 960.
11. Yu, W.; Fleming, M.; Mazsaroff, I.; Scoble, H.; Martin, S., in *Proceedings of the 42nd ASMS Conference on Mass Spectrometry and Allied Topics*, Chicago (1994) p. 442.
12. Cornett, D.S.; Peschke, M.; LaiHing, K.; Cheng, P.Y.; Willey, K.F. and Duncan, M.A., *Rev. Sci. Instr.* **63** (1992) 2177–2186.
13. Jia, W.J.; Ledingham, K.W.D.; Scott, C.T.J.; Kosmidis, C. Singhal, R.P., *Rapid Commun. Mass Spectrom.* **9** (1995) 761–766.
14. Cornish, T.J. and Cotter, R.J., *Rapid Commun. Mass Spectrom.* **8** (1994) 781–785.
15. Cordero, M.M.; Cornish, T.J.; Cotter, R.J.; Lys, I.A., *Rapid Commun. Mass Spectrom.* **9** (1995) 1356–1361.

9

Tandem TOF and Hybrid Instruments

For the analysis of biological samples, the tandem mass spectrometer provides two very important functions. The first mass analyzer (MS1), used primarily as a *mass filter*, performs the final step in the separation of complex mixtures that are often incompletely fractionated by chromatographic methods. The second mass analyzer (MS2) can then provide structural information for every component of the mixture by recording the product-ion mass spectra for each mass-selected precursor. Ideally, the ionization method used should provide only molecular ions, and the first mass analyzer should be capable of unit mass selection of each monoisotopic precursor. Ion activation is then provided in an intermediate region by collision-induced dissociation (CID), surface-induced dissociation (SID) or photodissociation (PD), and the products are analyzed as monoisotopic product-ion mass spectra. In this context, the PSD method for recording product ions on a MALDI–TOF mass spectrometer provides an imperfect approach to tandem mass-spectral measurements. Activation of ions during the ionization process (by increasing laser power) results in metastable fragmentation in regions of the instrument (during ion acceleration and in the reflectron) that are not refocused in the final spectra. In addition, higher laser power necessarily increases ion kinetic energy and decreases mass resolution. Finally, mass selection is generally accomplished by electrostatic gating in the first linear region (L_1) before the ions have been brought into full time focus by the reflectron, where its performance is further reduced by the increased kinetic energy resulting from higher laser power.

Nonetheless, we begin this chapter by briefly revisiting the tandem capabilities of PSD instruments. Following that, we describe several reports on the design and performance of tandem time-of-flight instruments utilizing two distinct TOF analyzers. These include linear tandem (TOF–TOF) mass spectrometers, and instruments (TOF–RTOF and RTOF–RTOF) incorporating one or two reflectrons. In addition, because of the physical limitations in achieving high performance (unit mass) ion gating, a number of investigators have proposed designs for hybrid instruments incorporating a double-focusing (EB) sector configuration for the first mass analyzer. Because MS1 is utilized as a mass filter, these EB–TOF instruments provide the same multichannel advantage for recording product-ion mass spectra as tandem time-of-flight mass spectrometers.

Mass Selection and Mixture Analysis in PSD Instruments

Figure 9.1 shows the source and linear regions of a PSD time-of-flight mass spectrometer reported by Cordero et al.[1] The electrostatic gating system is located close to the coaxial dual-channelplate detector, and the instrument utilizes a curved-field reflectron. Figure 9.2 shows the MALDI mass spectrum of a mixture of poly-ethylene glycol oligomers before and after application of the mass selection gate. The appearance of only minor amounts of oligomeric ions adjacent to the oligomer ion at m/z 925 indicates that the gate transmits ions with masses that are ±4% of the selected mass. Figure 9.3 shows the MALDI mass spectrum of an unfractionated tryptic digest of cytochrome C. The identity of the tryptic fragments can be identified from their molecular masses, including the trypsin autolysis peak: tryp [44–49]. After recording this mass spectrum, the laser power is raised slightly to induce metastable fragmentation, and gated spectra are obtained for each the tryptic fragments. Figure 9.4 shows the PSD mass spectra of the three largest fragments designated as cyt C [88–91], tryp [44–49] and cyt C [28–38]. These spectra were all obtained at full reflectron voltage,[2] requiring the acquisition of far fewer mass spectra than would be needed using conventional single- and dual-stage reflectrons. Nonetheless, the ability to select specific peptides from this mixture for product-ion analysis depends upon the fact that their molecular masses are very different.

Tandem Time-of-Flight Instruments

The introduction of MALDI as an effective ionization method for peptides, proteins, and other biopolymers has provided considerable motivation for the development of tandem time-of-flight instruments that would compete favorably with electrospray ionization, which enjoys easy compatability with tandem, triple quadru-pole mass spectrometers. Tandem TOF instruments would be considerably less expensive alternatives for routine sequencing of peptides obtained from enzymatic digests or from other biological mixtures.

TOF–TOF Instruments.   A simple design for a tandem (TOF–TOF) instrument that incorporates two drift regions separated by a collision chamber has been reported by Jardine et al.[3] As shown in Figure 9.5, the instrument uses fast-atom bombardment ionization. Because this is a continuous ionization technique, a deflection gating system is used to define the initial ion packet, while a second set of deflectors (located close to the collison chamber) is used for mass selection. In addition, the collision cell is raised to a voltage that is intermediate between ground and the source potential, so that product ions will be reaccelerated and will have velocities different from their precursors.

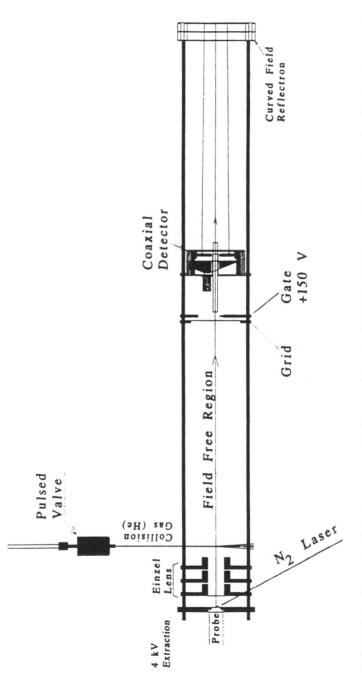

FIGURE 9.1 Schematic of the ion source and linear regions of a PSD time-of-flight mass spectrometer showing the electrostatic gating system and coaxial detector. (Reprinted with permission from reference 1).

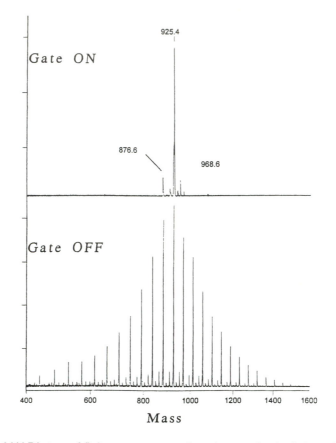

FIGURE 9.2 MALDI time-of-flight mass spectra of a mixture of polyethylene glycol oligomers (a) before and (b) after application of the mass selection gate. (Reprinted with permission from reference 1).

In this scheme, precursor ions have flight times (t_1) in the drift region (D_1) before the collision chamber equal to:

$$t_1 = \left(\frac{m_1}{2eV_a}\right)^{1/2} D_1$$

where V_a is the voltage on the source. The collision chamber is placed at a voltage V_c and bounded by grounded grids that provide rapid deceleration and reacceleration regions that are neglected in this calculation. As the precursor ions enter the collision chamber, they do so with energies equal to:

$$\frac{1}{2} m_1 v^2 = e(V_a - V_c)$$

Intact precursor ions emerging from the collision chamber are reaccelerated to final energies:

$$\frac{1}{2} m_1 v^2 = e(V_a - V_c) + e V_c = e V_a$$

and velocities:

$$v = \left(\frac{2 e V_a}{m_1}\right)^{1/2}$$

so that their flight times in the second drift region (D_2) are:

$$t_2 = \left(\frac{m_1}{2 e V_a}\right)^{1/2} D_2$$

In contrast, product ions leave the collision chamber with energies:

$$\frac{1}{2} m_2 v^2 = e \frac{m_2}{m_1} (V_a - V_c)$$

and are reaccelerated to final energies:

$$\frac{1}{2} m_2 v^2 = e \frac{m_2}{m_1} (V_a - V_c) + e V_c$$

and velocities:

$$v = \left[\frac{2e}{m_1}\left(V_a - V_c + \frac{m_1}{m_2} V_c\right)\right]^{1/2}$$

The flight time of the product ions in D_2 is then:

$$t_2 = \left(\frac{m_1}{2 e V_a}\right)^{1/2} \frac{D_2}{\left[1 + \frac{V_c}{V_a}\left(\frac{m_1}{m_2} - 1\right)\right]^{1/2}}$$

so that product ions arrive at different times than their precursors.

A more complete treatment that includes the times spent in the deceleration and reacceleration regions is given in the report by Jardine et al.[3]

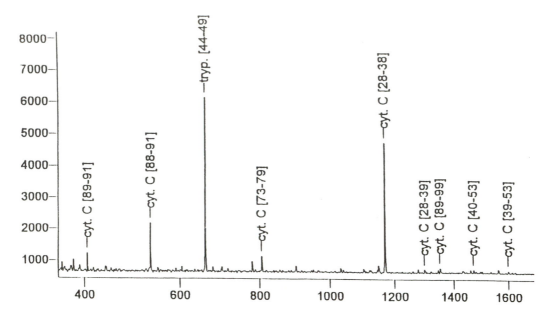

FIGURE 9.3 MALDI time-of-flight mass spectrum of an unfractionated tryptic digest of cytochrome C. (Reprinted with permission from reference 2).

In instruments in which the voltage on the collision cell (or photodissociation region) is raised, precursor ions will be recorded at flight times $(t_1 + t_2)$ equal to:

$$t = \left(\frac{m_1}{2eV_a}\right)^{1/2} \left[D_1 + D_2\right] \qquad [1]$$

while product ions will be observed at:

$$t = \left(\frac{m_1}{2eV_a}\right)^{1/2} \left[D_1 + \frac{D_2}{\left[1 + \frac{V_c}{V_a}\left(\frac{m_2}{m_1} - 1\right)\right]^{1/2}}\right] \qquad [2]$$

TOF–RTOF instruments.   An alternative approach to reacceleration uses a reflectron in MS2 to disperse the product-ion arrival times in the same manner as used for the post-source decay method. Additionally, the reflectron can be used in combination with reacceleration after the collision chamber. In the TOF–RTOF instrument reported by Cooks et al.[4] (Figure 9.6), ion activation is provided by surface-induced dissociation (SID), and product ions are dispersed and focused by a reflectron. The target surface is raised to a potential above the drift region, so that product ions are dispersed by reacceleration as well. The effect of this combination is to

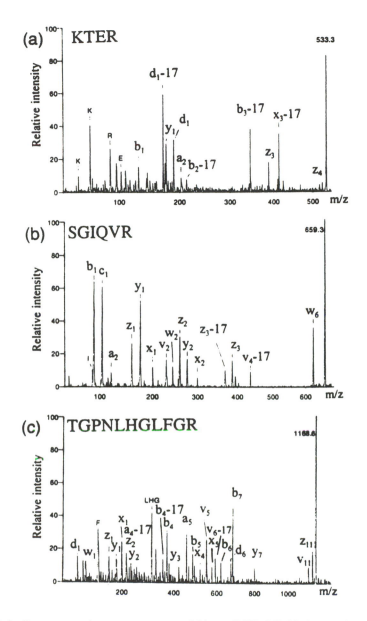

FIGURE 9.4   Post-source decay mass spectra of (a) cyt C [88–91], (b) the autolysis peak corresponding to tryp [44–49], and (c) cyt C [28–38]. (Reprinted with permission from reference 2).

reduce the range of product-ion energies entering the reflectron, resulting in focusing over a broader mass range for a particular reflectron voltage.

A tandem (TOF–RTOF) instrument has also been reported by Schlag et al.[5] in which MS1 is the drift region from the source to the space-focus plane, which in principle can be made any length using dual stage extraction, with longer drift lengths improving mass selectability. (Note that this situation is different from the simple PSD schemes described in the last chapter, where gating was carried out in a region

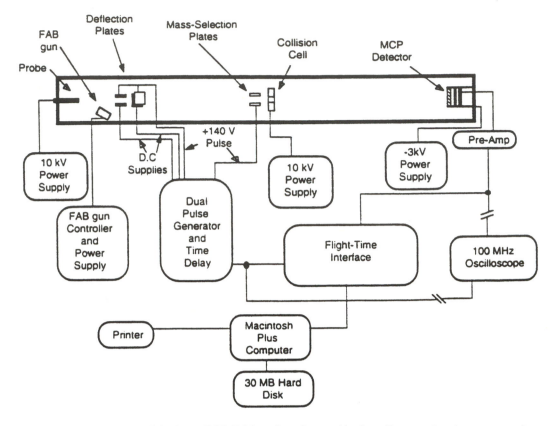

FIGURE 9.5   Schematic of the linear TOF–TOF tandem design of Jardine. (Reprinted with permission from reference 3).

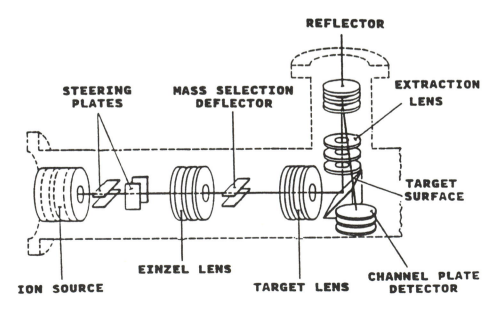

FIGURE 9.6   Schematic of a tandem TOF–RTOF instrument that employs surface-induced dissociation. (Reprinted with permission from reference 4).

in which the ions were not in focus). Ions are photodissociated at the space-focus plane, and can be reaccelerated prior to entering MS2, (in their instrument) a dual-stage reflectron mass analyzer. Again, as in the post-source decay method, the reflectron voltage must be stepped to focus different regions of the product-ion mass spectrum. However, fewer voltage steps are required when the ions are reaccelerated.

Reacceleration of ions following collision or photodissociation can be accomplished very simply by providing a flight-tube liner for the first drift region (D_1) at the intermediate voltage. In this case, the flight times, compared with those expressed in equations (1) and (2) for a collision cell at intermediate voltage, are only slightly different. In particular, the flight times in the first region become:

$$t_1 = \left(\frac{m_1}{2e\Delta V}\right)^{1/2}$$

where $\Delta V = V_a - V_c$, while the flight times in the second region remain the same as expressed in equations (1) and (2).

In either case, it is useful to express flight times in terms of the difference in arrival times for a product-ion mass (m_2) and its precursor (m_1). For an instrument using two linear analyzers this becomes:

$$t_b - t_a = \left(\frac{m_1}{2eV_a}\right)^{1/2}\left[\frac{D_2}{\left[1+\frac{V_c}{V_a}\left(\frac{m_2}{m_1}-1\right)\right]^{1/2}} - D_2\right]$$

In an instrument in which MS2 has a single-stage reflectron, the drift length (D_2) is replaced with two linear regions (L_1 and L_2) and the reflectron penetration depth (d'). The difference in flight times for product ions and their precursors in the linear regions ($L = L_1 + L_2$) is similar to that expressed above:

$$t_{b,L} - t_{a,L} = \left(\frac{m_1}{2eV_a}\right)^{1/2}\left[\frac{L}{\left[1+\frac{V_c}{V_a}\left(\frac{m_2}{m_1}-1\right)\right]^{1/2}} - L\right]$$

while the difference in flight times in the reflectron is more complex:

$$t_{b,R} - t_{a,R} = \left(\frac{m_1}{2eV_a}\right)^{1/2}4\left[\frac{d'}{\left[1+\frac{V_c}{V_a}\left(\frac{m_2}{m_1}-1\right)\right]^{1/2}} - d\right]$$

since the penetration depth (d') for a product ion is:

$$d' = \left[\frac{m_2}{m_1} + \left(1 - \frac{M_2}{m_1}\right)\frac{V_c}{V_a}\right] d$$

In TOF–RTOF instruments incorporating both reacceleration and a reflectron, the difference in arrival times for a product ion and its precursor is given by:

$$t_b - t_a = \left(\frac{m_1}{2eV_a}\right)^{1/2}\left[\frac{L}{\lambda} - L + \frac{d'}{\lambda} - d\right] \qquad [3]$$

where:

$$\lambda = \left[1 + \frac{V_c}{V_a}\left(\frac{m_2}{m_1} - 1\right)\right]^{1/2} \qquad [4]$$

and:

$$d' = \left[\frac{m_2}{m_1} + \left(1 - \frac{m_2}{m_1}\right)\frac{v_c}{V_a}\right] d \qquad [5]$$

While complex, equation (3) provides a suitable mass calibration equation using two or more known product-ion masses.

RTOF–RTOF Instruments.   The use of reflectron TOF mass analyzers for both MS1 and MS2 provides the opportunity to bring ions into full time focus at the point of mass selection. RTOF–RTOF instruments have been reported by Cornish and Cotter[6–8] and Enke et al.[9,10] using collision-induced dissociation (CID) and photodissociation, respectively. In the first instrument reported by Cotter and Cornish,[6] two dual-stage reflectrons were used and CID was provided by a pulsed valve (Figure 9.7a). Figure 9.8 shows the mass spectrum of the MALDI matrix compound: α-cyano-4-hydroxy-cinnamic acid, recorded through both reflectron mass analyzers, the mass spectrum following selection of the molecular ion, and the product-ion mass spectrum obtained when the collision gas valve is actuated. The broad peak appearing at approximately 16 μs in Figure 9.8c corresponds to unresolved low-mass ions reflected in the first stage of the reflectron. Figure 9.7b shows a later instrument[7] using two, single-stage reflectrons. Interestingly, extraordinarily high mass resolutions can be obtained by passing the ions through two reflectrons, even while the instrument has a total length less than 1 m. Figure 9.9 shows mass resolutions for the peptide gramicidin S (MW = 1142) and the laser dye rhodamine 6-G (MW = 429) of 11,230 and 8,983, respectively. Figure 9.10 shows the product-ion mass spectrum for methionine enkephalinamide obtained by mass selection of the molecular ion.

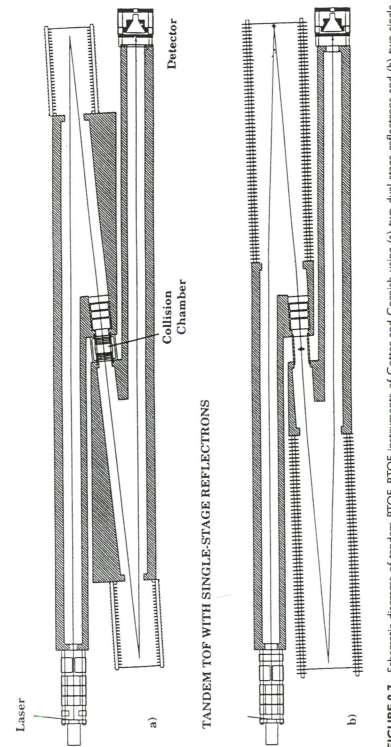

FIGURE 9.7 Schematic diagrams of tandem RTOF–RTOF instruments of Cotter and Cornish using (a) two dual-stage reflectrons and (b) two single-stage reflectrons. (Reprinted with permission from reference 7).

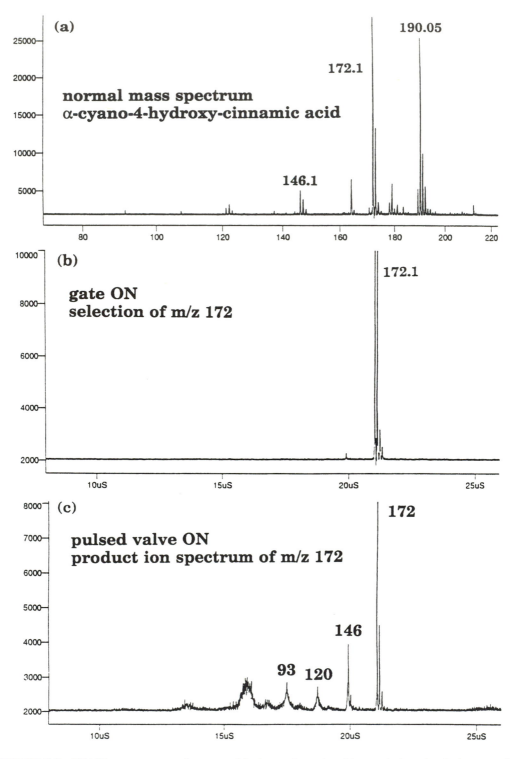

FIGURE 9.8   MALDI mass spectra of α-cyano-4-hydroxy-cinnamic acid recorded on the dual-stage reflectron tandem mass spectrometer. (a) Normal mass spectrum obtained through both reflectron analyzers, (b) mass spectrum after gating of the molecular ion, and (c) product-ion mass spectrum after activating the pulsed valve. (Reprinted with permission from reference 6).

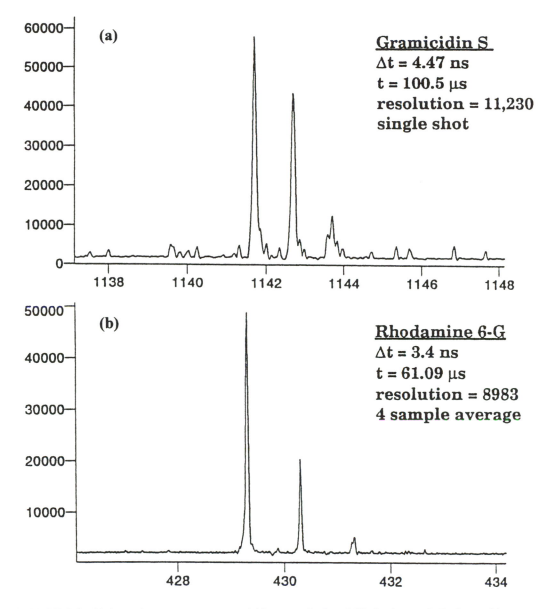

FIGURE 9.9   High resolution mass spectra of (a) gramicidin S and (b) rhodamine 6-G, obtained by passing the ions through two single-stage reflectrons.

While abundant fragmentation is observed, time-focusing (resolution) is poor using a single-stage reflectron at a single reflectron voltage.

In a more recent version of their tandem instrument (Figure 9.11), Cornish and Cotter used a curved-field reflectron in the second mass analyzer.[8] In this instrument the mass spectrum of caffeic acid, obtained by passing the ions through both reflectrons, revealed an extraordinary mass resolution of 1 part in 13,500 (Figure 9.12). In addition, the mass spectra in Figure 9.13 illustrate the fact that unit mass resolution

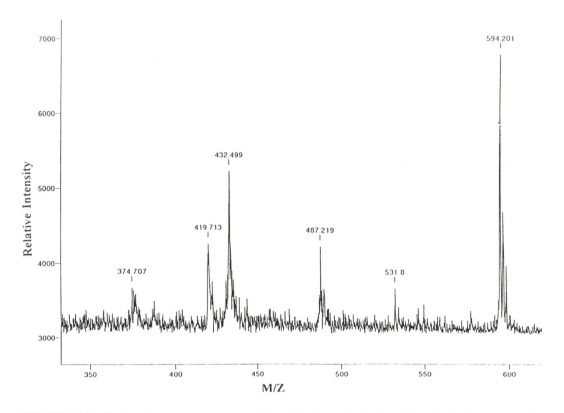

FIGURE 9.10 Product-ion mass spectrum of methionine enkephalinamide, obtained on a tandem (RTOF–RTOF) time-of-flight mass spectrometer with two single-stage reflectrons. (Reprinted with permission from reference 7).

can be obtained for product ions using the nonlinear reflectron at a single reflectron voltage.

This instrument has also been used to study the fragmentation behavior of fullerenes using different collision gases and collision gas pressures.[11] Figure 9.13 shows the mass spectrum of a fullerene sample containing both C_{60} and C_{70} and mass isolation of each of these species using the deflection gating system located before the collision chamber. Figures 9.14 and 9.15 show the product-ion mass spectra for C_{60} at different collision gas pressures, using helium and argon, respectively, as the collision gas. As might be expected, collision with the heavier gas leads to more extensive fragmentation at lower gas pressures.

Enke and co-workers[9] have also developed a tandem reflectron (RTOF–RTOF) mass spectrometer (Figure 9.16) in which ion activation is provided by photodissociation using the 193-nm wavelength of an ArF excimer laser. The ion dissociation region is floated to reduce the range of kinetic energies entering the second reflectron and, in their latest version, a nonlinear reflectron provides wide energy focusing.[10] The SIMION representation of their reflectron (Figure 9.17) shows some similarities with a two-stage reflectron, except that the first region is nonlinear, and there is a smooth transition between the regions of the reflectron. Figure 9.18 shows the

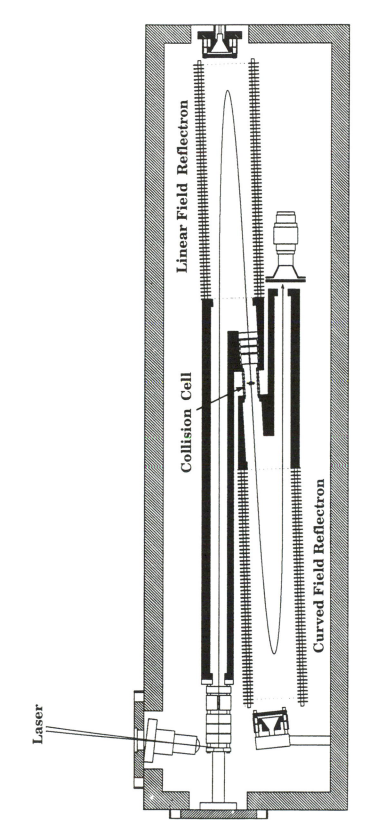

FIGURE 9.11 Schematic of a tandem (RTOF–RTOF) time-of-flight mass spectrometer in which the second reflectron is a curved-field reflectron. (Reprinted with permission from reference 8).

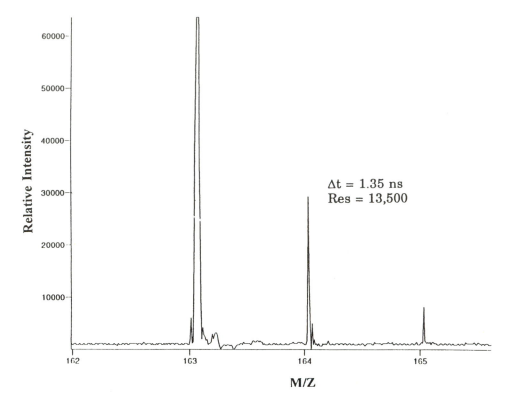

FIGURE 9.12   Molecular-ion region of the mass spectrum of caffeic acid, obtained on a tandem (RTOF–RTOF) time-of-flight mass spectrometer with a curved-field reflectron.

tandem photodissociation mass spectrum of the m/z 91 peak of toluene. The mass spectrum was calibrated using the peaks at m/z 39 and 65, and the equation:

$$t_a - t_b = \left(\frac{m_a}{\alpha + \beta}\right)^{1/2} - \left(\frac{m_b}{\alpha\dfrac{m_b}{m_a} + \beta}\right)^{1/2}$$ [6]

where t_a and t_b are the flight times of the precursor (m_a) and product (m_b) ions, respectively, and:

$$\alpha = \frac{2U_a}{L^2}$$ [7]

$$\beta = \frac{2U_b}{L^2}$$ [8]

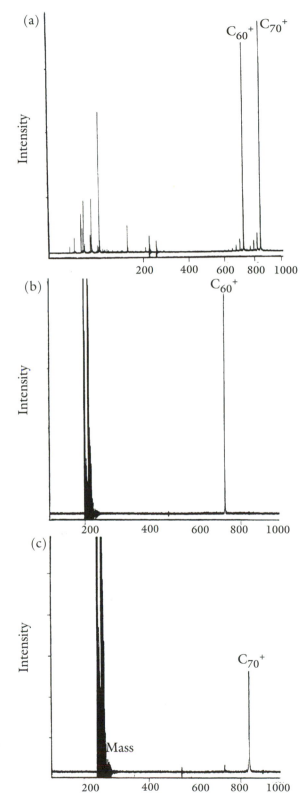

FIGURE 9.13 MALDI mass spectrum of a fullerene mixture, obtained on a tandem (RTOF–RTOF) time-of-flight mass spectrometer with a curved-field reflectron. (a) Normal mass spectrum, (b) mass spectrum after mass selection of the C_{60}^+ ion, and (c) mass spectrum after selection of the C_{70}^+ ion. (Reprinted with permission from reference 11).

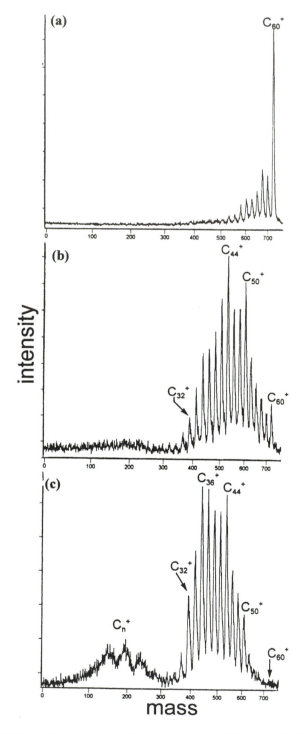

FIGURE 9.14   Collision-induced dissociation mass spectra of C_{60}^+, using helium as the target gas at (a) low [20%], (b) medium [80%] and (c) high [98%] attenuation. (Reprinted with permission from reference 11).

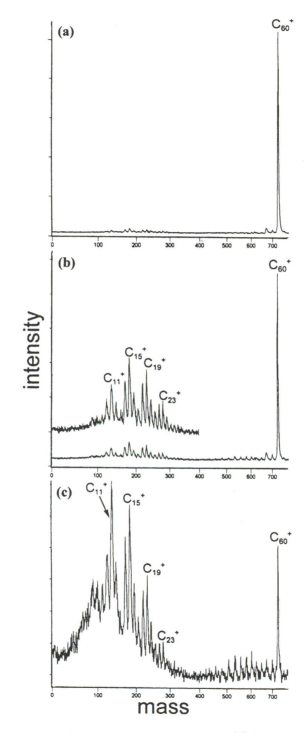

FIGURE 9.15 Collision-induced dissociation mass spectra of C_{60}^+, using argon as the target gas at (a) low, (b) medium, and (c) high attenuation. (Reprinted with permission from reference 11).

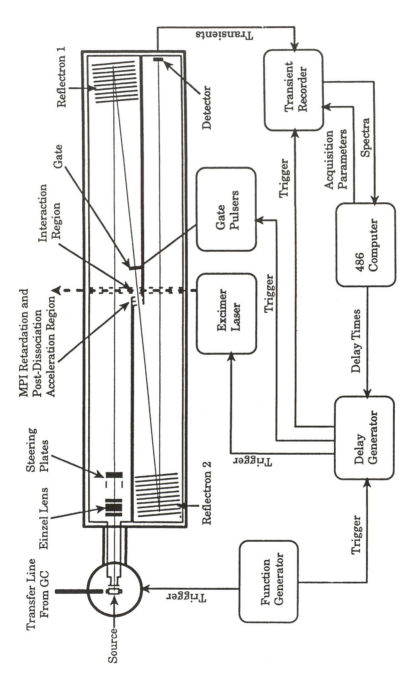

FIGURE 9.16   Tandem time-of-flight mass spectrometer, using photodissociation with an excimer laser. (Reprinted with permission from reference 9).

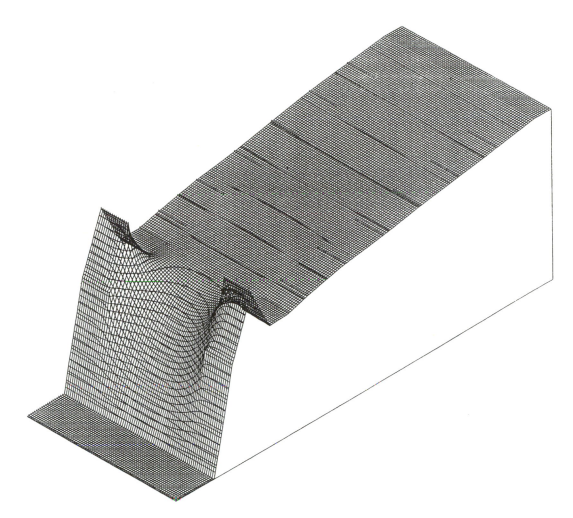

FIGURE 9.17   SIMION representation of the broad energy-focusing reflectron designed by Enke et al. (Reprinted with permission from reference 10).

U_a is the average kinetic energy of the precursor ions, and U_b is the energy gained by post-dissociation acceleration.

Hybrid (EB–TOF) Instruments

The use of a double-focusing sector mass analyzer as MS1 provides the opportunity for unit mass selection, and a number of designs have been proposed. In the instrument reported by Strobel and Russell,[12] ionization is provided by an excimer or nitrogen laser, or by a cesium ion gun (Figure 9.19). In the latter case, a deflector lens is used to provide the start pulse. The collision cell is floated to a voltage intermediate between the accelerating voltage and ground potential. This decelerates

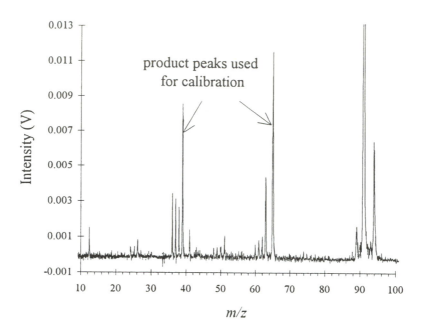

FIGURE 9.18 Tandem time-of-flight mass spectrum of the product ions formed by photodisso-ciation of the m/z 91 peak of toluene. (Reprinted with permission from reference 10).

the ions prior to entering the collision chamber, and provides additional kinetic energy to the product ions, enabling wider mass range focusing by the reflectron.

Orthogonal Extraction. When the sector mass analyzer passes single mass ions formed by continuous ionization methods (e.g., FAB or electrospray), the use of deflection techniques for forming the initial ion packet results in considerable loss of the available duty cycle. In addition, as we have seen, the combination of post-collisional acceleration with conventional single-stage and dual-stage reflectrons reduces, but does not eliminate, the need for stepping the reflectron voltage and acquiring multiple spectral segments. A design intended to address both of these problems was initially proposed by Clayton and Bateman in 1992.[13] In their scheme, shown in Figure 9.20), the kinetic energies of mass-selected ions from MS1 are reduced by deceleration lenses to 800 eV prior to entering the collision chamber, and the product ions are extracted at right angles to the direction of their velocity distributions. Because velocity distributions are minimal along the time-of-flight direction, reasonable focusing is possible using dual-stage (space-focusing) ion extraction into a linear time-of-flight mass analyzer. The reduction of initial kinetic energies to 800 eV reduces (but does not eliminate) the off-axis drift of ions that are recorded on a broad channelplate detector, and maintains sufficient translational energy for collisional activation. In addition, the extraction of ions from a large storage volume provides improvements in the duty cycle. At the same time, this scheme limits the range of collisional energies that can be utilized, although heavier collision gases can be employed to raise the relative collision energy in the center-of-mass frame. In addition, because of the considerable ion drift, this approach is not easily amenable

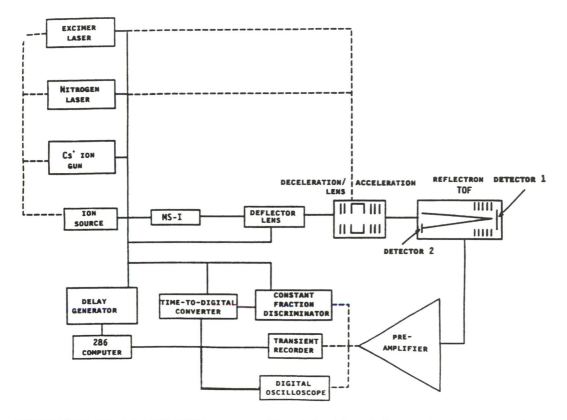

FIGURE 9.19   The hybrid EB–TOF instrument of Stroebel and Russell. (Reprinted with permission from reference 12).

to designs incorporating reflectrons to improve mass resolution. Figure 9.21 shows the tandem-sector orthogonal-TOF mass spectrum of melittin.[14]

Nonlinear Reflectrons.   Other approaches to hybrid sector–TOF instruments extract ions along the same axis into reflectrons having a wide energy-focusing range. In an instrument patented by Davis and Evans,[15] the initial ion packet is developed from a pulsed ion source using a combination of deflection lenses (D1) and an *ion buncher*. The latter is used to compress a pulse of mass-selected ions formed by the deflectors into a much shorter pulse, and consists of two lenses (*P1* and *P2*) that are normally at ground. When the initial pulse lies completely within the space between the two lenses, P1 is ramped to a positive voltage to provide an accelerating field that (essentially) space-focuses the ions at a point beyond the collision chamber, located at the entrance to the time-of-flight mass analyzer. The TOF analyzer is itself a quadratic reflectron (described in Chapter 3) which focuses the resultant kinetic energy spread formed at the space-focus point.

   An alternative approach proposed by Musselman et al.[16] utilizes a curved-field reflectron TOF analyzer as MS2. In their scheme, MS1 would be scanned (when continuous ionization is used), while the curved-field reflectron TOF samples the

MS1

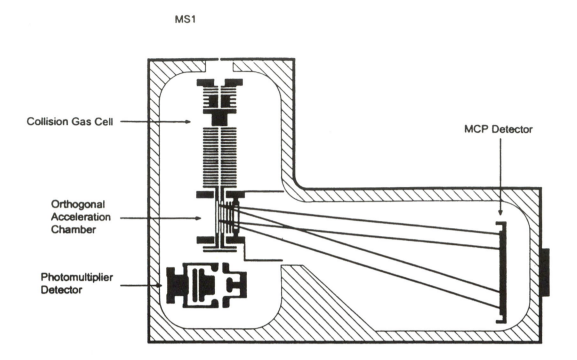

Collision Gas Cell

MCP Detector

Orthogonal
Acceleration
Chamber

Photomultiplier
Detector

FIGURE 9.20 The orthogonal extraction EB–TOF instrument of Clayton and Bateman. (Reprinted with permission from reference 14).

ion current over many cycles for each mass spectral peak. Integration of the total ion current in each time-of-flight cycle produces the normal mass spectrum, while the digitized transients provide real-time product-ion analysis, since the reflectron voltage does not need to be scanned.

Conclusions

At this point, the tandem time-of-flight story is far from complete. Instruments incorporating sector mass analyzers as MS1 have a clear advantage for unit mass selection, but will necessarily be somewhat more expensive than current sector mass spectrometers. Improvements in nonlinear reflectron design will undoubtable be key to the success of both hybrid (EB–TOF) and tandem (TOF–TOF) instruments that provide the high sensitivity, multichannel recording advantage. In addition, the possibilites for utilizing the MS–MS capabilities of combined ion trap–TOF instruments (as suggested in Chapter 7) have not been fully exploited.

Applications

As we leave this chapter, the last on instrumentation, we focus on some of the unique accomplishments of the current time-of-flight technology in solving biological

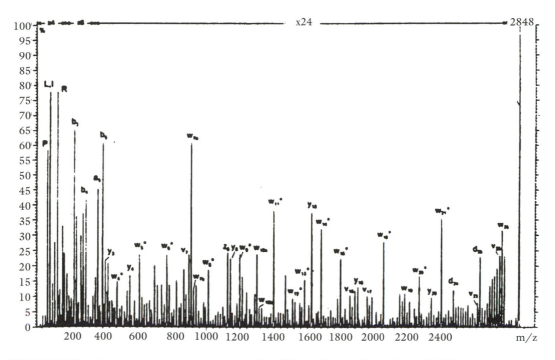

FIGURE 9.21 Product-ion mass spectrum for the MH⁺ ion from melittin (*m/z* = 2847.5), resulting from 800-eV (35-eV center-of-mass) collisions with xenon and orthogonal-extraction TOF analysis. (Reprinted with permission from reference 14).

problems involving proteins, peptides, and glycopeptides (Chapter 10) and in sequencing the human genome (Chapter 11). We will then conclude (Chapter 12) with a discussion of the future prospects for time-of-flight mass spectrometry.

References

1. Cordero, M.M.; Cornish, T.J.; Cotter, R.J.; Lys, I.A., *Rapid Commun. Mass Spectrom.* **9** (1995) 1356–1361.
2. Fabris, D.; Vestling, M.M.; Cordero, M.M.; Doroshenko, V.M.; Cotter, R.J.; Fenselau, C., *Rapid Commun. Mass Spectrom.* **9** (1995) 1051–1055.
3. Jardine, D.R.; Morgan, J.; Alderdice, D.S. and Derrick, P.J., *Org. Mass Spectrom.* **27** (1992) 1077–1083.
4. Schey, K.L.; Cooks, R.G.; Kraft, A.; Grix, R.; Wollnik, H., *Int. J. Mass Spectrom. Ion Processes* **94** (1989) 11–14.
5. Boesl, U.; Weinkauf, R.; Schlag, E.W., *Int. J. Mass Spectrom. Ion Processes* **112** (1992) 121–166.
6. Cotter, R.J.; Cornish, T.J., *Anal. Chem.* **65** (1993) 1043–1047.
7. Cornish, T.J.; Cotter, R.J., *Org. Mass Spectrom.* **28** (1993) 1129–1134.
8. Cornish, T.J.; Cotter, R.J., *Rapid Commun. Mass Spectrom.* **8** (1994) 781–785.
9. Seeterlin, M.A.; Vlasak, P.R.; Beussman, D.J.; McLane, R.D.; Enke, C.G., *J. Am. Chem. Soc.* **4** (1993) 751–754.
10. Beussman, D.J.; Vlasak, P.R.; McLane, R.D.; Seeterlin, M.A.; Enke, C.G., *Anal. Chem.* **67** (1995) 3952–3957.

11. Cordero, M.M.; Cornish, T.J.; Cotter, R.J., *J. Am. Soc. Mass Spectrom.* **7** (1996) 590–597.

12. Strobel, F.H.; Preston, L.M.; Washburn, K.S.; Russell, D.H., *Anal. Chem.* **64** (1992) 754.

13. Clayton, E.; Bateman, R.H., *Rapid Commun. Mass Spectrom.* **6** (1992) 719–720.

14. Bateman, R.H.; Green, H.R.; Scott, G.; Clayton, E., *Rapid Commun. Mass Spectrom.* **9** (1995) 1227–1233.

15. Davis, S.C.; Evans, S., European Patent Application EP 0 551 999 A1 (1993)

16. Cotter, R.J.; Cornish, T.J.; Musselman, B., *Rapid Commun. Mass Spectrom.* **8** (1994) 339–340.

10

Peptides, Proteins, and Glycoproteins

Although the amino acid sequences of peptides can be effectively determined by Edman degradation, the strength of mass spectrometry in biological research is undoubtably the ability to provide structural information for a wide range of biological polymers that include peptides, carbohydrates, lipids, and oligonucleotides. This is particularly important given the fact that most expressed proteins also carry oligosaccharides, phosphate, sulfate, and/or fatty acyl modifications. Mass spectrometry can also be used to provide detailed structural analysis of other mixed biopolymers, including glycolipids, phosphorylated lipopolysaccharides, sulfated oligosaccharides, phospholipids, and heme proteins. In fact, the structural questions that mass spectrometry can address range broadly from the determination of disulfide bond positions in proteins to the elucidation of noncovalent protein-protein, enzyme-substrate, and protein-DNA interactions.

Not surprisingly, the analytical strategies for the mass spectral analysis of proteins (peptides), carbohydrates, and oligonucleotides are quite similar. Peptide mapping provides a set of peptide fragments from a protein that can be compared to a protein database or used to locate a particular post-translational modification. Peptide maps are generally obtained from the mass spectra of proteins that have been digested chemically or with enzymes (endoproteinases) specific for one or more amino acids, while comparable structural information for carbohydrates and oligonucleotides can be obtained using specific endoglycosidases or endonucleases (restriction enzymes), respectively. Similarly, sequencing of peptides, oligosaccharides, and oligonucleotides can be carried out using *ladder* approaches that include exopeptidase, exoglycosidase, or exonuclease digestion (respectively), or chemical hydrolysis prior to mass spectral analysis. In addition, ladder methods for sequencing oligonucleotides that synthesize oligonucleotides in the presence of chain-terminating didexoy nucleotides have a counterpart in peptide sequencing in the building of combinatorial libraries. Finally, structural information for biological molecules can be obtained from fragmentation in the mass spectrometer.

In this chapter, we describe the mapping, ladder, and fragmentation approaches for the structural analysis of peptides, proteins, and glycoproteins, in some cases using examples from the literature. In Chapter 11, we focus specifically on oligonucleotides, which present a particular challenge with respect to ionization efficiency, mass range, and mass resolution, and the need for large-scale DNA sequencing for sequencing the human genome.

Strategies Based upon Molecular Weight Measurements

Although the last three chapters have described methods (post-source decay, tandem time-of-flight, and delayed extraction) that can provide structural information from fragmentation occurring in the mass spectrometer, the most common approaches have been those that are based upon molecular weight measurements, generally in combination with chemical or enzymatic digestion. As has been noted earlier (Chapter 2), prompt extraction of ions generally precludes the observation of fragmentation, which (for large molecules) occurs over a longer time frame. While such post-source fragmentation can indeed be observed in reflectron instruments, simple (linear) instruments currently provide the highest sensitivity.

Molecular weight strategies for the structural analysis of proteins and peptides are summarized in Figure 10.1. The simplest measurement is (of course) the molecular weight of an intact protein (or glycoprotein), where the precision is considerably higher than can be obtained by gel electrophoresis. Reduction of a protein containing disulfide bonds using dithiothreitol increases the molecular mass by 2 Da for each disulfide bond. For high-molecular-weight proteins, larger mass increases can be observed by converting the free sulfhydryl groups of cysteines to their S-carboxymethyl derivatives using iodoacetamide.[1] Both native and reduced protein can then be subjected to chemical and/or enzymatic digestion. Cyanogen bromide (CNBr) cleaves at methionine residues, which are generally infrequent, resulting in large fragments. Thus, a common approach is to separate CNBr fragments by HPLC and carry out additional, enzymatic digestion reactions. Subsequent mass-spectral analysis of the unfractionated products then provides a peptide map, and comparison of maps derived from native and reduced proteins locates those peptides linked by disulfide bonds. Additionally, purification of specific peptide fractions provides opportunities for amino acid sequencing and location of protein modifications using mass spectrometry. When amino acid sequences are known, direct mass-spectral analysis of a glycopeptide (for example) may be used to assess carbohydrate heterogeneity, while the presence of a phosphate group may be determined by conversion of the phosphoserine residue to S-ethyl cysteine.[2] Finally, treatment with carboxypeptidases or aminopeptidases provides a set of nested peptides having a common amino or carboxy terminus, respectively. Mass spectra of these digests reveal a set of molecular-ion peaks, with the differences in mass between adjacent peaks corresponding to specific amino acid residues. In addition, differences in mass corresponding (for example) to a phosphoserine residue can provide the exact location of phosphorylation.

Molecular Weights of Peptides and Proteins.   Traditionally, organic mass spectroscopists have utilized peaks corresponding to monoisotopic ion species (containing only ^{12}C, ^{1}H, ^{14}N, ^{16}O, ^{32}S, etc.) to determine molecular or fragment masses (Table 10.1). While their relative abundances are low, the contribution to the molecular-ion mass from one or several ^{13}C, ^{2}H, ^{15}N, and ^{18}O isotopes increases for peptides and proteins containing a large number of carbon, hydrogen, nitrogen, and oxygen

MOLECULAR WEIGHT STRATEGY

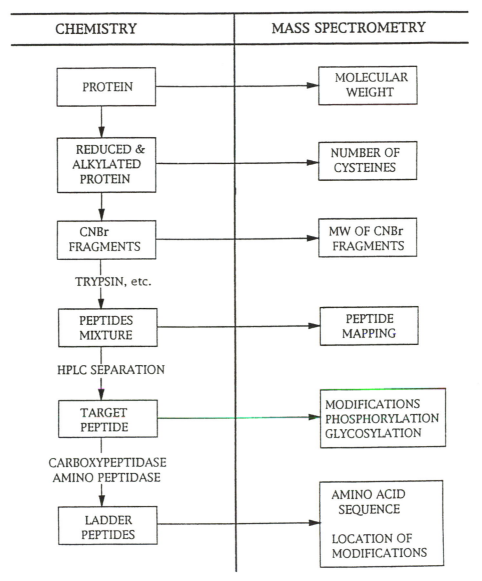

| CHEMISTRY | MASS SPECTROMETRY |

PROTEIN → MOLECULAR WEIGHT

REDUCED & ALKYLATED PROTEIN → NUMBER OF CYSTEINES

CNBr FRAGMENTS → MW OF CNBr FRAGMENTS

TRYPSIN, etc.

PEPTIDES MIXTURE → PEPTIDE MAPPING

HPLC SEPARATION

TARGET PEPTIDE → MODIFICATIONS PHOSPHORYLATION GLYCOSYLATION

CARBOXYPEPTIDASE AMINO PEPTIDASE

LADDER PEPTIDES → AMINO ACID SEQUENCE

LOCATION OF MODIFICATIONS

FIGURE 10.1 Molecular weight strategies for protein and peptide analysis by time-of-flight mass spectrometry.

atoms.[3] For example, the protonated molecular ion for the peptide glucagon has an elemental formula $C_{153}H_{225}N_{42}O_{50}S$ that results in a monoisotopic ion peak in its mass spectrum corresponding to m/z 3482.6. However, as shown in Figure 10.2a, it is not the most intense peak. The second peak at m/z 3483.6 results from contributions from three unresolved isotopic species:

TABLE 10.1 Isotopic abundances and masses of elements encountered in peptides and other biological molecules.

Element	Nominal mass	Exact mass	Percent abundance	Average mass
C	12	12.00000	98.9	
	13	13.00335	1.1	12.01115
H	1	1.00783	99.98	
	2	2.0140	0.02	1.008665
O	16	15.99491	99.8	
	18	17.9992	0.20	15.994
N	14	14.00307	99.63	
	15	15.00011	0.37	14.0067
S	32	31.97207	95.0	
	33	32.97146	0.76	
	34	33.96786	4.22	32.064

---

1. $^{12}C_{152}{}^{13}C_1{}^1H_{225}{}^{14}N_{42}{}^{16}O_{50}{}^{32}S_1$, $m/z = 3483.6109$, RA $= 153 \times 1.1\% = 170\%$
2. $^{12}C_{153}{}^1H_{224}{}^2H_1{}^{14}N_{42}{}^{16}O_{50}{}^{32}S_1$, $m/z = 3483.6137$, RA $= 225 \times 0.02\% = 4.5\%$
3. $^{12}C_{153}{}^1H_{225}{}^{14}N_{41}{}^{15}N_1{}^{16}O_{50}{}^{32}S_1$, $m/z = 3483.6048$, RA $= 42 \times 0.37\% = 15.5\%$

whose abundances (RA) relative to those of the monoisotopic species are determined by multiplying the isotopic abundances for each element with the number of that element in the molecular formula. Thus, the intensity of the second peak will be 190% of the intensity of the monoisotopic peak. The third peak at m/z 3484.6 results from six major (>1%) isotopic species and is the most intense peak. In Figure 10.2a, the isotopic distribution for the protonated molecular ion of glucagon is shown as it would appear at a mass resolution of 8000 (FWHM), with relative intensities normalized to the most intense peak.

Figure 10.2b shows the molecular-ion peak for glucagon as it would be observed at a mass resolution of 1200 (FWHM). In this case the isotopic contributions are not resolved, but it is possible to determine the average molecular-ion mass (m/z 3484.8) from the peak centroid. Such measurements are common in time-of-flight mass spectra when the mass exceeds the resolution, and generally have a mass accuracy in the range of 0.1% to 0.01%.

Calculating Molecular Weights and Mass Balancing. When the amino acid sequence is known or can be deduced from the sequence of the gene encoding the protein, the molecular weight can be calculated using amino acid residue masses and adding the mass of water. Monoisotopic and average residue masses for the common amino acids are listed in Table 10.2 (along with their single- and three-letter codes) and correspond to the molecular weights of free amino acids minus the mass of water lost in forming peptide bonds. Using this table, the average molecular weight for the peptide *DAEFR* (a tryptic fragment of β-amyloid) can be calculated to be 636.7 Da:

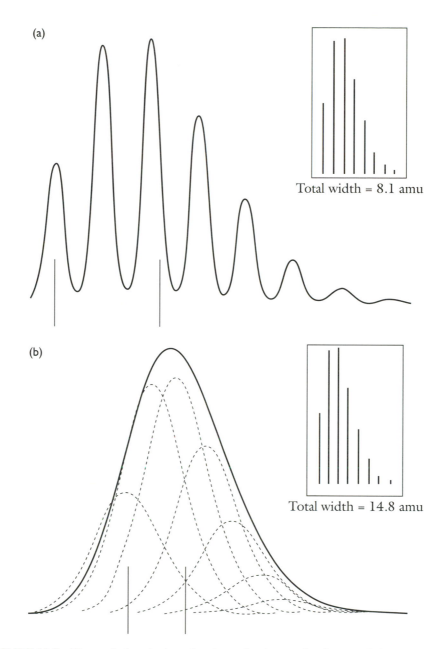

FIGURE 10.2 Theoretical peak shape for the molecular-ion distribution of glucagon calculated (a) at unit mass resolution, and (b) at a mass resolution of 1200 (FWHM).

$$\begin{array}{ccccccc} 1 & 115.1 & 71.1 & 129.1 & 147.2 & 156.2 & 17 \\ \text{H} - \text{Asp} & - & \text{Ala} - & \text{Glu} - & \text{Phe} - & \text{Arg} - & \text{OH} \end{array}$$

When molecular weights can be calculated, mass analysis of the intact peptides and proteins can be used to reveal point mutations, differences in post-translational modifications to native and recombinant proteins, or the amount of molecular mass

---

TABLE 10.2   Masses of the 20 common amino acid "residues".

Amino acid	Three-letter code	Single-letter code	Nominal mass	Monoisotopic mass	Average mass
Alanine	Ala	*A*	71	71.037	71.079
Arginine	Arg	*R*	156	156.101	156.188
Asparagine	Asn	*N*	114	114.043	114.104
Aspartic acid	Asp	*D*	115	115.027	115.089
Cysteine	Cys	*C*	103	103.009	103.143
Glutamic acid	Glu	*E*	129	129.043	129.116
Glutamine	Gln	*Q*	128	128.059	128.131
Glycine	Gly	*G*	57	57.021	57.052
Histidine	His	*H*	137	137.059	137.141
Isoleucine	Ile	*I*	113	113.084	113.160
Leucine	Leu	*L*	113	113.084	113.160
Lysine	Lys	*K*	128	128.095	128.175
Methionine	Met	*M*	131	131.040	131.197
Phenylalanine	Phe	*F*	147	147.068	147.177
Proline	Pro	*P*	97	97.053	97.117
Serine	Ser	*S*	87	87.032	87.078
Threonine	Thr	*T*	101	101.048	101.105
Tryptophan	Trp	*W*	186	186.079	186.214
Tyrosine	Tyr	*Y*	163	163.063	163.176
Valine	Val	*V*	99	99.068	99.133

---

due to carbohydrate in glycoproteins. For proteins whose amino acid sequence is unknown, or only partially known, analysis of the intact protein provides an estimate of the molecular weight that can insure that subsequent mass analyses of tryptic (or other enzymatic) fragments and amino acid sequencing accounts for the entire protein. This approach is known as *mass balancing*, and requires that the sum of the molecular weights of n enzymatic fragments, minus $(n\text{-}1)$ water molecules, add up to the mass of the molecular weight of the intact peptide or protein. For example, the β-amyloid peptide (βA_{1-40}):

DAEFRHDSGYEVHHQKLVFFAEDVGSNKGAIIGLMVGGVV

has a molecular weight of 4330.4 Da. This peptide, which is found as senile plaques in the brain of Alzheimer's disease patients, includes a string of hydrophobic residues (*VGGVV*) on the carboxy terminus and is not easily solubilized in aqueous solutions. In a recent study, peptides obtained from Alzheimer's disease brain were digested with trypsin (which cleaves at arginine and lysine residues) and mapped using plasma desorption mass spectrometry.[4] Four major peaks were observed at *m/z* 637.8,

1086.1, 1326.7, and 1337.1. Assuming that these peaks correspond to MH⁺ ions, the mass balance can be checked from the molecular weights:

$$636.8 + 1085.1 + 1325.7 + 1336.1 - 3(18) = 4329.7$$

to verify that all of the peptides have been observed. An additional (smaller) peak was also observed at *m/z* 522.4, leading to the assignments:

AEFR	MW = 521.6	βA_{2-5}
DAEFR	MW = 636.7	βA_{1-5}
HDSGYEVHHQK	MW = 1336.5	βA_{6-16}
LVFFAEDVGSNK	MW = 1325.7	βA_{17-28}
GAIIGLMVGGVV	MW = 1085.5	βA_{29-40}

Interestingly, the βA_{1-40} form of the peptide is also found in normal brain, while longer forms that include additional hydrophobic residues on the carboxy terminus have been implicated in Alzheimer's disease. Thus, further digestion of this same peptide sample with CNBr (which cleaves at methionine residues) resulted in three additional peaks at *m/z* 431.1, 614.2, and 626.0, corresponding to the peptides:[4]

VGGVV	MW = 429.6	βA_{36-40}
VGGVVIA	MW = 613.8	βA_{36-42}
GAIIGLM	MW = 625.8	βA_{29-35}

revealing the longer βA_{1-42} peptide:

DAEFRHDSGYEVHHQKLVFFAEDVGSNKGAIIGLMVGGVVIA

Molecular Weights of Reduced Proteins.   In describing molecular weight strategies, we noted the importance of obtaining molecular weight measurements for both native and reduced proteins. In 1992, Chait and co-workers[5] reported the structure of the fully processed spider toxin protein w-Agatoxin IA (Figure 10.3). Initially, they obtained a molecular weight of 7791.4 Da for the mature toxin using plasma desorption mass spectrometry, which was not consistent with the mass calculated from the amino acid sequence for the major chain of the toxin. Subsequent treatment with dithiothreitol and mass analysis by matrix-assisted laser desorption/ionization revealed a molecular weight of 7495.3 Da, from which they were able to conclude that the mature toxin included a minor (3 amino acid) chain, linked to the major chain by a disulfide bond.

Peptide Mapping

Tryptic mapping provides a convenient means for comparing recombinant proteins and native proteins. In this technique, recombinant proteins are digested with trypsin, fractionated on reversed-phase HPLC, and the retention times of their tryptic fragments compared with those from native proteins.[6] Since the amino acid sequence is known and should be the same as that of the native protein, differences in retention

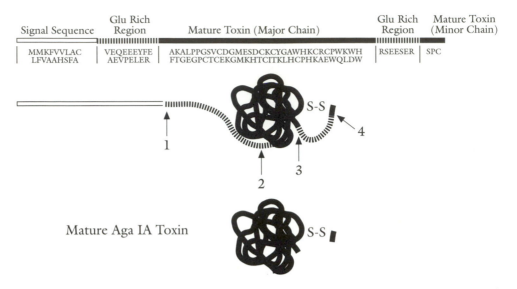

FIGURE 10.3   Processing scheme for the w-agatoxin IA precursor to the mature toxin. Reprinted with permission from reference 5.

time will quickly reveal any differences in post-translational processing. At the same time, a change in retention time does not provide direct information on the nature of the post-translational modification, while a change in the mass could reveal that processing of the protein involved cleavage at a different site, glycoslyation, or fatty acylation. Thus, Naylor et al.[7] in 1986 proposed an approach known as *FAB mapping*, in which unfractionated tryptic digests were analyzed by fast-atom bombardment mass spectrometry to reveal any changes in mass. This approach is now used with other ionization techniques (PDMS, MALDI, and ESI) and is generally referred to as peptide mapping or mass mapping. In addition, mass spectroscopists have generally expanded their repertoire of proteolytic enzymes used for producing such maps, and included chemical digestion methods as well. These are described here, and can be used for both mapping experiments and for isolation of specific peptides (using HPLC) for additional structural studies.

Chemical Methods.   The most common chemical method for digesting proteins uses cyanogen bromide (CNBr) to cleave at methionine residues. CNBr cleaves the amide bond on the C-terminal side of methionine:

$$\text{⸺NH} - \text{CH} - \overset{\overset{\displaystyle O}{\|}}{\text{C}} - \text{NH} - \overset{\overset{\displaystyle R}{|}}{\text{CH}} - \text{CO⸺}$$
$$|$$
$$\text{CH}_2$$
$$|$$
$$\text{CH}_2$$
$$|$$
$$\text{S}$$
$$|$$
$$\text{CH}_2$$

forming a terminal homoserine:

$$\text{........ NH}-\underset{\underset{\displaystyle CH_2-CH_2-OH}{|}}{\text{CH}}-\overset{\overset{\displaystyle O}{\|}}{\text{C}}-\text{OH}$$

or homoserine lactone residue:

$$\text{........ NH}-\underset{\underset{\underset{\displaystyle CH_2}{\diagdown}}{\displaystyle CH_2}}{\text{CH}}-\text{C}\underset{\diagdown O}{\overset{\diagup\!\diagup O}{}}$$

This change in mass (from that of methionine) must be accounted for. For example, the βA_{29-35} fragment *GAIIGLM* (described above) would be expected to have a mass of 673.9 Da. However, when generated by cleavage with CNBr the mass of a peptide containing homoserine would be 30.1 Da less (643.8), while the peptide containing homoserine lactone would be 48.1 Da less (625.8). Thus, the peptide observed in that experiment contained the homoserine lactone.

Other chemical reagents include *N*-bromosuccinimide (NBS), *N*-chlorosuccinimide (NCS), *o*-iodobenzoic acid, and DMSO/HCl/HBr (which cleave at tryptophan residues); hydroxylamine (which cleaves the amide bond between asparagine and glycine); and 70% formic acid (which cleaves the amide bond between aspartic acid and proline). Of these, NBS and NCS are the most commonly used. Like CNBr they tend to produce large fragments, which can be separated and subjected to secondary digestion with proteolytic enzymes. In addition, incomplete digestion results in overlapping fragments, which can be used to determine the order of the fragments.

Enzymatic Methods.   Trypsin is the most commonly used proteolytic enzyme and cleaves the amide bond C-terminal to arginine and lysine residues. In many cases, peptides will contain a number of adjacent arginine or lysine residues, resulting in single amino acid fragments that make mass-balancing techniques difficult. In such cases, it is often useful to use Endoproteinase-Lys C (which cleaves at lysines) or Endoproteinase-Arg C (which cleaves at arginines) to reduce the number of single amino acid fragments, and establish the order of the tryptic fragments. *S. aureus* V8 cleaves at glutamic acid and aspartic acid residues, while Endoproteinase-Glu C cleaves at glutamic acid. Both of these enzymes provide an alternative map to trypsin that can help to establish overlaps in sequence. Chymotrypsin cleaves at phenylalanine, tyrosine, tryptophan, histidine, leucine, and alanine residues, and can be used to distinguish leucine and isoleucine residues, while pepsin cleaves at phenylalanine and leucine residues. Two enzymes which cleave the amide bond N-terminal to a specific residue are Endoproteinase-Asp N (which cleaves at aspartic acid) and thermolysin (which cleaves at leucine, isoleucine, methionine, phenylalanine, and tryptophan residues).[8]

In Chapter 4 we described a number of examples using plasma desorption mass spectrometry to map enzymatic digests. However, continuing an earlier example from this chapter, plasma desorption mass spectrometric analysis of amyloid peptides (βA_{1-40} and βA_{1-42}) digested with pepsin resulted in peaks at m/z 561.1, 613.1, 746.1, 1492.2, 1999.2, 2199.9, and 2315.2 that could be assigned to the peptide fragments:[4]

DAEFRHDSGYEVHHQKLVF	MW = 2314.8	βA_{1-19}
AEFRHDSGYEVHHQKLVF	MW = 2199.4	βA_{2-19}
FRHDSGYEVHHQKLVF	MW = 1999.4	βA_{4-19}
FAEDVGSNKGAIIGL	MW = 1490.7	βA_{20-34}
MVGGVV	MW = 560.8	βA_{35-40}
MVGGVVIA	MW = 745.1	βA_{35-42}
VGGVVIA	MW = 613.8	βA_{36-42}

As expected, cleavages occurred at leucine and phenylalanine residues, although incomplete cleavage at the Phe_4 residue resulted in the fragments βA_{1-19} and βA_{2-19}. And, in contrast to the tryptic digest, cleavage at the Leu_{34} residue resulted in both βA_{35-40} and βA_{35-42} fragments and revealed the presence of full-length amyloid peptide. In addition, formation of the βA_{36-42} fragment suggested additional cleavage at the Met_{35} residue. Unanticipated cleavages are common when working with proteolytic enzymes. In our own work, we have found that Endo-Arg C also cleaves the amide bond between adjacent lysines,[9] that prolonged digestion with Endo-Arg C or Endo-Lys C results in cleavage of both lysine and arginine residues, and that Endo-Glu C cleaves the amide bonds on either side of glutamic acid when preceded by an alanine residue.[2] In general, the extent of digestion as well as its specificity depend upon the enzyme/substrate ratio, the length of the digestion period, pH, the buffers used, temperature, and the accessibility of residues to the enzyme. In addition, the proteolytic enzymes used to digest peptides will also be cleaved, resulting in additional peaks in the mass spectrum. These autolysis peaks can be identified by mass analysis of a blank having the same enzyme concentrations, pH, buffers, and digestion times, but leaving out the peptide substrate. In general the extent of autolysis will depend upon the enzyme/substrate ratio, which should be about 1:50.

Immunoprecipitation Methods.   Zhao and Chait[10] described a *linear epitope mapping* strategy that compares the mass maps of an enzymatic digest and the specific peptide fragments that can be isolated by immunoprecipitation with a monoclonal antibody raised against the intact protein or peptide. In one example, a MALDI mass spectrum (map) of the Lys-C digest of this peptide (Figure 10.4a) revealed peaks corresponding to the fragments:

GIGAVLKVLTTGLPALISWIKRKRQQ	MW = 2847.9	(1–26)
GIGAVLKVLTTGLPALISWIKRK	MW = 2435.4	(1–23)
VLTTGLPALISWIKRKRQQ	MW = 2209.0	(8–26)
VLTTGLPALISWIKRK	MW = 1796.5	(8–23)

This peptide mixture was then immunoprecipitated using a monoclonal antibody raised against melittin. The unbound peptides were washed away, and the bound peptides released and analyzed by MALDI mass spectrometry. As shown in Figure 10.4b, the monoclonal antibody bound only the peptides including residues

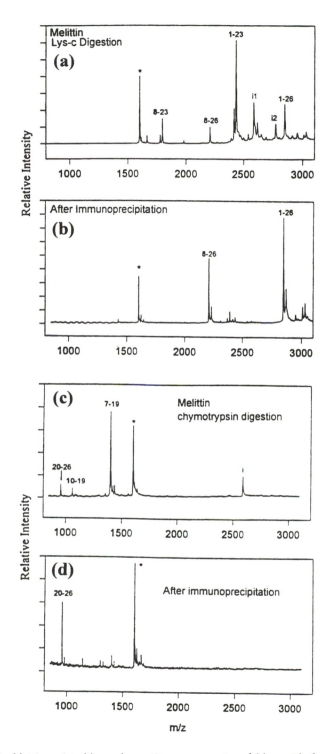

FIGURE 10.4 Matrix-assisted laser desorption mass spectra of (a) peptide fragments produced by endoproteinase Lys-C digestion of melittin, (b) Endo Lys-C fragments isolated by immunoprecipitation with mab 83144, (c) chymotrypsin fragments of melittin, and (d) chymotrypsin fragments of melittin immunoprecipitated with antimelittin antibody mab 83144. Peaks labeled with an asterisk are dynorphan A_{1-13} added as internal mass calibrant, and peaks labeled I1, I2, and I are impurities. Reprinted with permission from reference 10.

1–26 and 8–26. A second digestion was carried out using chymotrypsin. Mapping of this digest revealed peaks corresponding to the fragments:

IKRKRQQ	MW = 956.3	(20–26)
KVLTTGLPALISW	MW = 1398.9	(10–19)
TTGLPALISW	MW = 1058.4	(7–19)

that are shown in Figure 10.4c. Immunoprecipitation of this digestion mixture with monoclonal antibody and MALDI mass spectrometry revealed a single peak (Figure 10.4d) corresponding to the residues 20–26, which located the epitope within this region.

Peptide Ladder Sequencing

Trypsin, chymotrypsin, Endo-Arg C, and other proteolytic enzymes, which we have described above; are endoproteinases (or endopeptidases). Endoproteinases and chemical reagents such as CNBr, NBS, and NCS cleave at specific amino acid residues, producing peptide fragments that can be mapped or fractionated for amino acid sequencing using chemical, enzymatic, or mass spectrometric methods. Edman degradation is the most common method for sequencing peptide fragments. However, a number of mass spectrometric methods for amino acid sequencing have also been developed, using either ladder or fragmentation approaches. The ladder approaches (which we consider here) involve progressive, single-amino-acid degradation of peptides from either the carboxy or amino terminus, resulting in a mixture of *nested* (carboxy or amino *co-terminal*) peptides that are analyzed by mass spectrometry. The amino acid sequence is then obtained from the mass differences between a series of molecular-ion peaks appearing in the spectrum.

Peptide ladders can be produced by exopeptidases (enzymes which catalyze hydrolysis of peptide bonds from either the carboxy or amino terminus), by Edman degradation reagents, or by chemical hydrolysis.

Carboxy and Aminopeptidases. Carboxypeptidases are exopeptidases that produce a ladder mixture of peptides having a common amino terminus. Continuing our previous example, the mass spectrum of a carboxypeptidase digest of the β-amyloid tryptic fragment *DAEFR* would produce a series of molecular ion peaks corresponding to:

$$\begin{array}{cccccccc} 1 & 115.1 & 71.1 & 129.1 & 147.2 & 156.2 & 17 & 1 \\ H & Asp & Ala & Glu & Phe & Arg & OH & H^+ \end{array}$$
H — Asp — Ala — Glu — Phe — Arg — OH — H^+ *M/Z* = 637.7

$$\begin{array}{ccccccc} 1 & 115.1 & 71.1 & 129.1 & 147.2 & 17 & 1 \\ H & Asp & Ala & Glu & Phe & OH & H^+ \end{array}$$
H — Asp — Ala — Glu — Phe — OH — H^+ *M/Z* = 481.5

$$\begin{array}{cccccc} 1 & 115.1 & 71.1 & 129.1 & 17 & 1 \\ H & Asp & Ala & Glu & OH & H^+ \end{array}$$
H — Asp — Ala — Glu — OH — H^+ *M/Z* = 334.3

$$\begin{array}{ccccc} 1 & 115.1 & 71.1 & 17 & 1 \\ H & Asp & Ala & OH & H^+ \end{array}$$
H — Asp — Ala — OH — H^+ *M/Z* = 205.2

$$\begin{array}{cccc} 1 & 115.1 & 17 & 1 \\ H & Asp & OH & H^+ \end{array}$$
H — Asp — OH — H^+ *M/Z* = 134.1

TABLE 10.3 Chemical reagents and proteolytic enzymes.

Chemical reagents	Cleavage sites
Cyanogen bromide (CNBr)	After *M*
BNPS-skatole or DMSO+HCl	After *W*
Acid hydrolysis	*D/P* then random

Endopeptidases	Cleavage sites
Trypsin	After *K/R*
Endoproteinase Lys-C	After *K*
Endoproteinase Glu-C	After *E*
Endoproteinase Asp-N	Before *D*
Endoproteinase Arg-C	After *R*
Chymotrypsin	After *F/W/Y/L*
Pepsin	After *F/W/Y/L*
Thermolysin	Before *L/I/M/F/W*

Exopeptidases	Selectivity
Carboxypeptidase A	Stops at *R/PX*, sometimes at *G/S/D/E*
Carboxypeptidase B	Cleaves at *R/K*
Carboxypeptidase P	Cleaves *PX/D/E*, sometimes stops at *S/G*
Carboxypeptidase Y	Cleaves at *PX/E*, sometimes stops at *K/R/S/G*
Aminopeptidase M	Nonspecific
Leucine aminopeptidase	Stops at or near *K/R/P*

Figure 10.5a depicts the mass spectrum that would be expected from a *perfect ladder*, that is, one in which all of the peptides are formed in equal abundances and have equivalent ionization efficiencies. However, commercially available carboxypeptidases have a wide range of specificities for hydrolysis of amide bonds adjacent to particular amino acid residues.[8,11] As shown in Table 10.3, carboxypeptidases will tend to stop at specific residues, or will cleave rapidly so that the mass difference between two peaks in the mass spectrum corresponds to two residues, whose order cannot be determined. Thus, it is important to carry out digestions using more than one carboxypeptidase, either separately or as a cocktail. In addition, it is also best to carry out a number of timed digestion reactions using several aliquots of the peptide solution, adding carboxypeptidase solution to each of these, and quenching the reactions at different times (from 10 to 120 s) by adding the matrix solution. These reactions can be carried out directly on the sample probe. Figure 10.6 shows the mass spectra of a series of timed digestion reactions of ACTH fragment 7–38, using carboxypeptidase Y.[12]

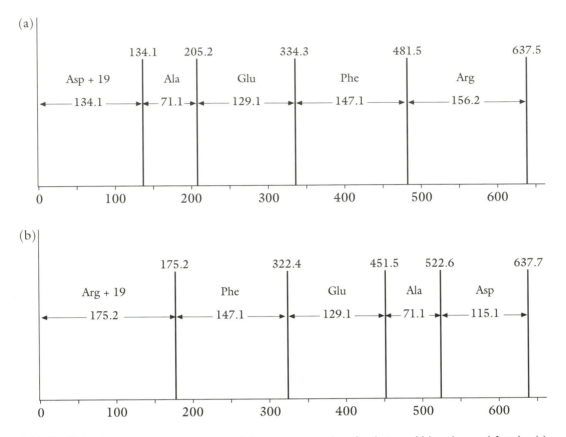

FIGURE 10.5 Histogram representation of the mass spectral peaks that would be observed for the (a) carboxypeptidase and (b) aminopeptidase digestion of the peptide *DAEFR* forming a "perfect" ladder mixture.

Aminopeptidases generally require somewhat longer reaction times (60 to 300 s). Commercially available amino peptidases include aminopeptidase M and leucine aminopeptidase. Again, digestion of the β-amyloid peptide fragment *DAEFR* using aminopeptidase would be expected to yield the molecular-ion species:

$$\underset{\text{H}}{\overset{1}{}} - \underset{\text{Asp}}{\overset{115.1}{}} - \underset{\text{Ala}}{\overset{71.1}{}} - \underset{\text{Glu}}{\overset{129.1}{}} - \underset{\text{Phe}}{\overset{147.2}{}} - \underset{\text{Arg}}{\overset{156.2}{}} - \underset{\text{OH}}{\overset{17}{}} - \underset{}{\overset{1}{\text{H}^+}} \qquad M/Z = 637.7$$

$$\underset{\text{H}}{\overset{1}{}} - \underset{\text{Ala}}{\overset{71.1}{}} - \underset{\text{Glu}}{\overset{129.1}{}} - \underset{\text{Phe}}{\overset{147.2}{}} - \underset{\text{Arg}}{\overset{156.2}{}} - \underset{\text{OH}}{\overset{17}{}} - \underset{}{\overset{1}{\text{H}^+}} \qquad M/Z = 522.6$$

$$\underset{\text{H}}{\overset{1}{}} - \underset{\text{Glu}}{\overset{129.1}{}} - \underset{\text{Phe}}{\overset{147.2}{}} - \underset{\text{Arg}}{\overset{156.2}{}} - \underset{\text{OH}}{\overset{17}{}} - \underset{}{\overset{1}{\text{H}^+}} \qquad M/Z = 451.5$$

$$\underset{\text{H}}{\overset{1}{}} - \underset{\text{Phe}}{\overset{147.2}{}} - \underset{\text{Arg}}{\overset{156.2}{}} - \underset{\text{OH}}{\overset{17}{}} - \underset{}{\overset{1}{\text{H}^+}} \qquad M/Z = 322.4$$

$$\underset{\text{H}}{\overset{1}{}} - \underset{\text{Arg}}{\overset{156.2}{}} - \underset{\text{OH}}{\overset{17}{}} - \underset{}{\overset{1}{\text{H}^+}} \qquad M/Z = 175.2$$

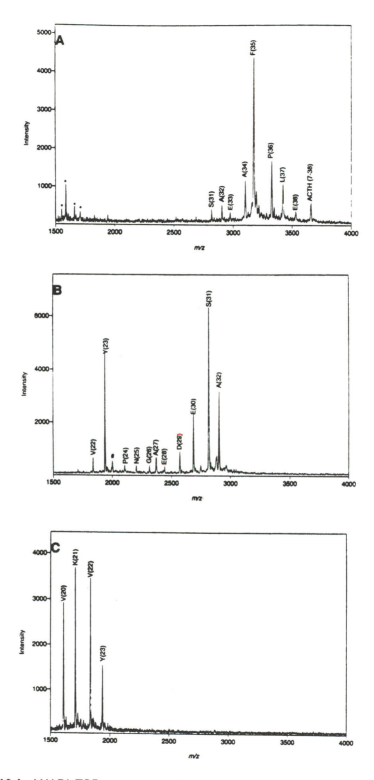

FIGURE 10.6 MALDI–TOF mass spectra of the (a) 1-min, (b) 5-min, and (c) 25-min aliquots from a time-dependent CPY digestion of ACTH 7–28 fragment (*FRWGKPVGKKRRPVKVYPNGAEDE-SAEAFPLE*). Reprinted with permission from reference 12.

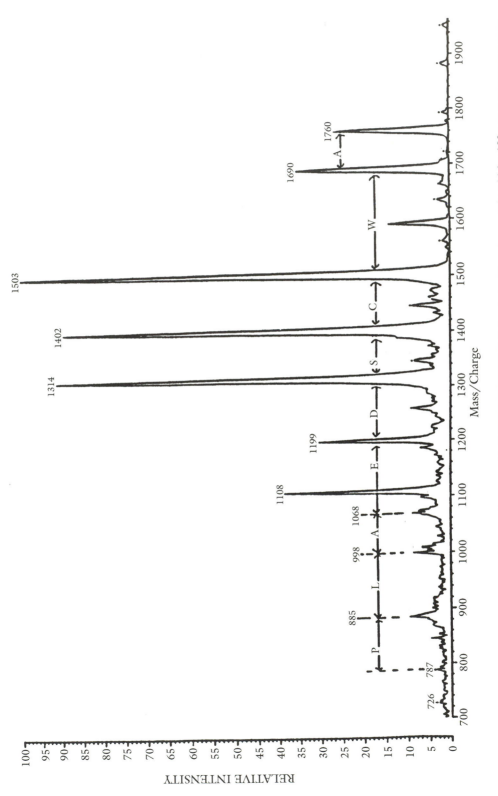

FIGURE 10.7 MALDI–TOF mass spectrum of the synthetic peptide AWCSDEALPPGSPRCDG digested with aminopeptidase M for 180 s at room temperature.

where the perfect ladder mass spectrum is shown in Figure 10.5b. Figure 10.7 shows a mass spectrum obtained for an aminopeptidase M digest of the synthetic peptide *AWCSDEALPPGSPRCDG*, which reveals nine amino acids from the amino terminus.[13] The peak at *m/z* 1108 is from a peptide added as a mass calibrant after digestion.

Sequencing of a Class I Peptide Antigen.   An example of these sequencing strategies is shown in Figure 10.8. As background to this example, we note that cells routinely degrade proteins in the cytosol into (generally) 9-amino-acid peptides that are then presented on cell surfaces bound to a transmembrane protein, known as the multiple histocompatibility (MHC) molecule, for recognition by cytotoxic T lymphocytes (CTL) or *killer* T cells. Peptides derived from endogenous proteins produced by tumor cells or by cells infected with a virus are not recognized as *self* peptides by CTL, which then lyse and destroy the cells. This Class I restricted recognition differs from the antigen–antibody arm of the immune system in which B-lymphocytes recognize epitopes of intact proteins or glycoproteins from exogenous invaders such as bacteria, since it recognizes peptides only when bound to MHC. An additional arm of the immune system (known as Class II) processes both exogenous proteins and endogenous proteins secreted from cells in special antigen presenting cells (APC) for recognition of MHC-bound peptides by *helper* T cells that respond to non-self peptides by secreting lymphokines that stimulate the production of CTL and B-lymphocytes.

Recently, DeCloux and co-workers[14] isolated a peptide bound to a novel murine Class IB MHC molecule known as Qa-1. Figure 10.8a shows the reversed-phase HPLC purification of the peptide in which the chromatographic fractions are monitored by a Qa-1 specific CTL assay that releases chromium upon cell lysis. The mass spectrum of bioactive fraction 45, to which ammonium citrate was added to enhance desorption, yielded a single peptide observed as its MH$^+$ and MNH$_4^+$ ions (insert). Figure 10.8b shows the mass spectrum of the aminopeptidase M digest of the bioactive fraction. This spectrum reveals the sequence of the first six amino acids from the amino terminus, although the order of the first two (alanine and methionine) cannot be established. The carboxypeptidase P digest of fraction 45 is shown in Figure 10.8c. The first three amino acids on the carboxy terminus are leucine or isoleucine, which have the same mass and cannot be distinguished. The next two amino acids (tyrosine and arginine) observed, provide an overlap with the aminopeptidase results. An aliquot of the bioactive fraction was also digested with chymotrypsin, which (as noted above) cleaves at leucines, but not isoleucines. In Figure 10.8d, the peak at *m/z* 915.0 was used to establish the terminal amino acid as alanine, while the other peaks were consistent with leucine as the seventh and eighth residues in this nine-amino-acid peptide. Leucine could not be distinguished from isoleucine as the carboxy terminal residue, so that the final sequence was determined to be: *AMAPRTLL(L/I)*.

In Situ Digestion and Desorption from Membranes.   While the electrospray ionization technique enjoys a considerable advantage as an online chromatographic (HPLC) detector, the direct desorption of proteins and peptides from electroblotted membranes using MALDI provides an equally important link to one-dimensional and two-dimensional gel electrophoresis. In such a scheme the laser beam would

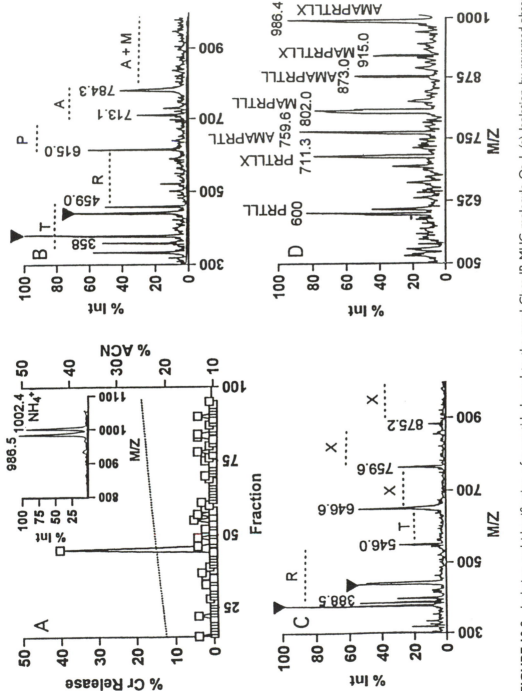

FIGURE 10.8 Isolation and identification of a peptide bound to the novel Class IB MHC molecule Qa-1. (a) Isolation by reversed-phase HPLC and identification of the bioactive fraction 45 by CTL assay. Insert is the MALDI–TOF mass spectrum of fraction 45. MALDI–TOF mass spectra of (b) aminopeptidase M, (c) carboxypeptidase P, and (d) chymotrypsin digests of fraction 45. Reprinted with permission from reference 14.

scan the entire two dimensional surface, initially producing molecular weight values that are more accurate than those obtained from the rf values. Additionally, because MALDI is nondestructive, the membranes containing proteins could be removed from the mass spectrometer, and specific spots or the entire surface digested with trypsin, reanalyzed by mass spectrometry, and compared with a protein database to rapidly identify all of the known proteins. While (at this time) time-of-flight mass spectrometers have not been designed specifically to accommodate large two-dimensional surfaces, it has been demonstrated that membranes commonly used to electroblot proteins are in fact excellent surfaces for desorbing proteins and peptides, and for carrying out in situ digestion reactions.

Specifically, Hillenkamp and co-workers[15] separated a mixture of proteins (hen egg-white lysozyme, soybean trypsin inhibitor, bovine carbonic anhydrase, hen egg-white ovalbumin, bovine serum albumin, and rabbit muscle phosphorylase) ranging from 14.4 kDa to 97.4 kDa using one-dimensional SDS-PAGE (sodium dodecyl sulfonate–polyacrylamide gel electrophoresis), electroblotted these onto a poly(vinylidene fluoride) (PVDF) membrane, and stained the membrane with Coomassie Blue to locate the bands. The bands were then cut from the PVDF membrane, attached to the sample probe tip, and analyzed by MALDI time-of-flight mass spectrometry. As shown in Figure 10.9, they compared the results using a UV (frequency quadrupled Nd:YAG) laser at 355 nm and an IR (Er:YAG) laser at 2.94 μm and observed better desorption efficiency (sensitivity) and mass resolution using the IR wavelength. They also noted that the commonly used Coomassie Blue stain adversely effected mass accuracy, and suggested the use of parallel (stained and unstained) lanes to locate unstained proteins for mass analysis.

Reliable identification of a known protein from a database can generally be obtained from the masses of only three or four of its enzymatic fragments. For example, Henzel et al.[16] separated a crude protein extract of *Escherichia coli* using two-dimensional electrophoresis (isoelectric focusing in the first dimension and electrophoresis by SDS/8–18% polyacrylamide gel in the second dimension). The proteins were then electroblotted onto a PVDF membrane (Figure 10.10a), and stained with Coomassie Blue. Each of the labeled spots (1–10) were cut from the membrane, alkylated, and digested in situ with dithiothreitol and trypsin, respectively, and mass analyzed using a pulsed nitrogen (337 nm) laser. The mass spectrum of the tryptic digest of spot 1 is shown in Figure 10.10b. Interrogation of the protein database identified this protein as cysteine synthase A (MW = 34,358.1 Da), identifying the following peaks:

ALGANLVLTEGAK	MH+ = 1257.8 Da	Δm = −0.3 Da
NIVVILPSSGER	MH+ = 1248.1 Da	Δm = −0.4 Da
VIGITNEEAISTAR	MH+ = 1475.5 Da	Δm = +0.9 Da
IQGIGAGFIPANLDLK	MH+ = 1627.4 Da	Δm = −0.5 Da
IFEDNSLTIGHTPLVR	MH+ = 1812.8 Da	Δm = −0.3 Da
not identified	MH+ = 1941.9 Da	

using a mass tolerance of 4 Da. All 10 of the labeled peaks were identified using this approach, and the identifications were confirmed by N-terminal (Edman) sequencing.

Taking this approach a step further, Vestling and Fenselau[17] digested bovine adenosine deaminase with BNPS-skatole, which cleaves at tryptophan residues and converts the terminal tryptophan to an oxolactone:

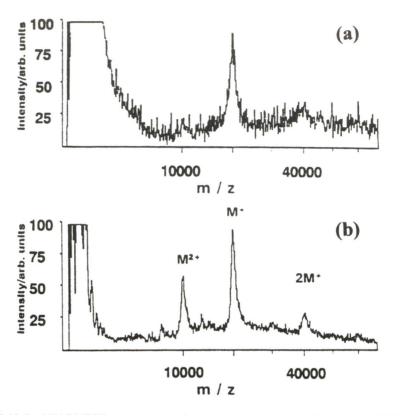

FIGURE 10.9 MALDI–TOF mass spectra of soybean trypsin inhibitor blotted onto PVDF membrane, using (a) UV laser at 355 nm and (b) IR laser at 2.94 μ. Matrix is 2,5-dihydroxy benzoic acid in ethanol. Reprinted with permission from reference 15.

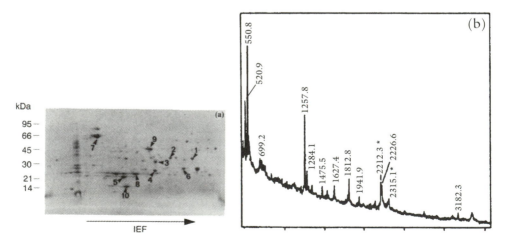

FIGURE 10.10 (a) *E. coli* proteins separated by isoelectric focusing in a pH 4–6 precast gel in the first dimension and by electrophoresis in an SDS/8–18% polyacrylamide gel in the second dimension, blotted onto a PVDF membrane and stained with Coomassie® brilliant blue. (b) MALDI–TOF mass spectrum of in situ tryptic digest of spot 1, identified as cysteine synthase A. Reprinted with permission from reference 16.

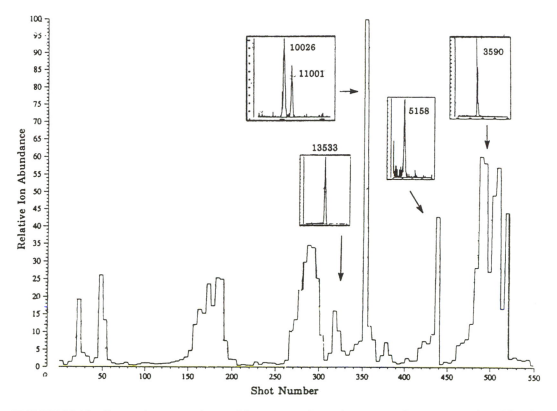

FIGURE 10.11 Electropherogram detected by scanning laser desorption of proteins produced by the reaction of BNPS-skatole and bovine adenosine deaminase, separated on polyacrylamide and electroblotted onto a PVDF membrane. Reprinted with permission from reference 17.

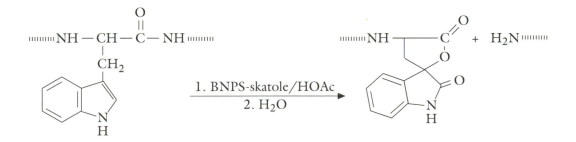

The peptide products were then separated by SDS-PAGE, and electroblotted onto a PVDF membrane. However, in this case the peptides were *not* stained. Instead, the entire separation lane was introduced to the mass spectrometer, and scanned by a 337-nm pulsed nitrogen laser, which produced the mass electrophoretograph (from the total ion current between m/z 500 and 3000) shown in Figure 10.11, and located the peptide bands. This "densitometer" plot shows that the separation of these peaks is well preserved after the addition of matrix, and that staining (which can interfere with the mass measurement) is not required.

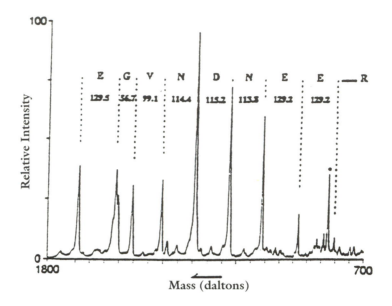

FIGURE 10.12 MALDI–TOF mass spectrum of the 14-residue [Glu^i]fibrinopeptide following eight cycles of manual ladder-generating Edman chemistry. Reprinted with permission from reference 20.

Edman Degradation Methods. Although the first reports of ladder sequencing of peptides by mass spectrometry utilized carboxypeptidases,[18,19] a number of other approaches have been used. Most notably, Chait et al.[20] have used a modified Edman approach in which the peptide (or its ladder products in later cycles) is reacted with a mixture of phenylisothiocyanate (PITC) + 5% V/V phenylisocyanate (PIC). In each cycle, PIC reacts with the αNH$_2$ group of 5% of the peptides to yield αNH$_2$-phenylcarbamyl (PC) peptides that are protected from further removal of amino acids. At the same time, PTIC reacts with the remaining peptides to form PTC-peptide derivatives from which the N-terminal amino acid can be removed by reaction with trifluoroacetic acid (TFA). The resultant truncated peptides can then be reacted with the PITC–PIC mixture in further cycles. At the end of the last cycle, a single MALDI mass spectrum is obtained for the mixture of PC-peptides, as shown in Figure 10.12. This approach can (of course) provide only N-terminal sequencing, but has the advantage (with respect to endopeptidases) that it does not particularly discriminate for specific residues. It also has an advantage over normal Edman methods, in that it can reveal modifications (such as phosphorylation, sulfation, or glycosylation) to amino acid residues. This is so because mass-spectrometric detection measures mass differences between the remaining PC-peptides, while conventional Edman sequencing detects the leaving residue. At the same time, this approach does not provide a real alternative to Edman sequencing (since it requires a gas-phase sequencer) and, because it involves chemical reactions carried out over many cycles, is more time-consuming than endopeptidase reactions which can be carried out within a few minutes.

Hydrolysis Methods. Vorm and Roepstorff[21] utilized a number of hydrolyzing reagents, including acetic acid, trifluoroacetic acid (TFA), phosphoric acid, sulphuric acid, and hydrochloric acid (HCl) to generate peptide ladders, and found HCl to be the most reliable. Figure 10.13 shows the mass spectrum of the hydrolyzed peptide

EVRFMVSEIPEA, which provided amino acid sequencing of the first eight residues from the carboxy terminus. They noted that while a simple, chemical approach, the relatively harsh sample treatment makes it unsuitable to localize the position of amidated residues, oxidized residues present prior to hydrolysis, or acid-labile post-translational modifications.

Synthetic Methods: Combinatorial Libraries.   Combinatorial libraries are large arrays of peptide sequences that can be screened for high affinity binding to receptors, enzymes or antibodies, and can therefore be used in the design of peptide-based pharmaceuticals. One approach involves the generation of *support-bound* libraries, in which a single sequence is bound to each of a large number of resin beads, each of which contains sufficient peptide to enable amino acid sequencing. In this case, beads which show high affinity binding to a particular receptor can be isolated, their peptides released, and then sequenced.

Keough et al.[22] have developed a *partial termination* synthetic method for building support-bound combinatorial libraries that enables amino acid sequencing of peptides using a mass spectrometric ladder approach. In their approach, a linker peptide (H$_2$N-*XXXRM*-) is attached to the resin beads (●) prior to synthesis of the library. The methionine residue provides a convenient cleavage point for release of the peptides using CNBr, the arginine residue provides a fixed positive charge, and the additional residues (*XXX*) ensure that the masses of all peptides in the ladder exceed 500 Da, so that they can be distinguished from the ion signal arising from matrix and other low-mass ions. In the first synthetic cycle, the resin-bound peptides (H$_2$N-*XXXRM*-●) are divided into 20 aliquots, and added to a reaction vessel containing 1 of the 20 normal amino acids (90%) and a capping reagent (CAP), acetyl-D,L alanine (10%). This first cycle then generates the support-bound sequences:

$$\text{H}_2\text{N-A}_1 XXXRM\text{-●}$$
$$\text{CAP-}XXXRM\text{-●}$$

where A$_1$ is the first additional amino acid. The beads from all 20 aliquots are then pooled, and redivided into 20 aliquots for the second cycle which produces:

$$\text{H}_2\text{N-A}_2\text{A}_1 XXXRM\text{-●}$$
$$\text{CAP-A}_1 XXXRM\text{-●}$$
$$\text{CAP-}XXXRM\text{-●}$$

This procedure is carried out for the desired number of cycles, and the final array of beads is screened for binding with a particular receptor. Beads that exhibit high-affinity binding can be isolated and reacted with CNBr to produce a set of ladder peptides:

$$\text{H}_2\text{N-A}_n\ldots\ldots\text{A}_2\text{A}_1 XXXRB$$
$$-$$
$$-$$
$$\text{CAP-A}_2\text{A}_1 XXXRB$$
$$\text{CAP-A}_2 XXXRB$$

where *B* represents the conversion of methionine to homoserine lactone. Figure 10.14 shows the ladder mass spectrum of a bead containing the sequence

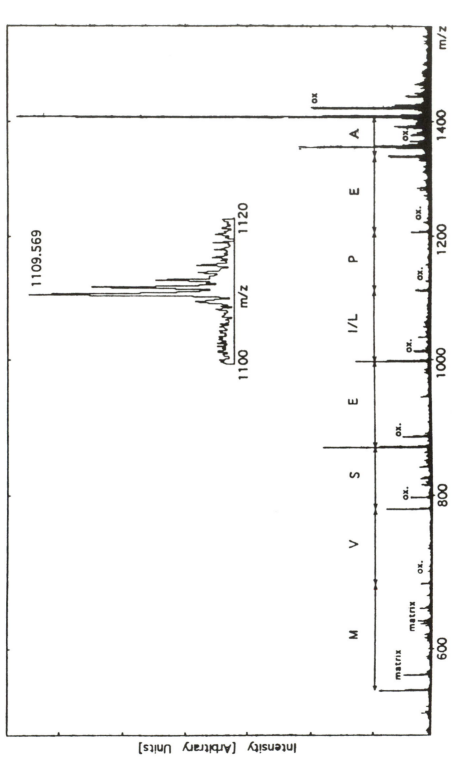

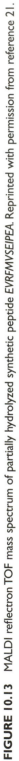

FIGURE 10.13   MALDI reflectron TOF mass spectrum of partially hydrolyzed synthetic peptide *EVRFMVSEIPEA*. Reprinted with permission from reference 21.

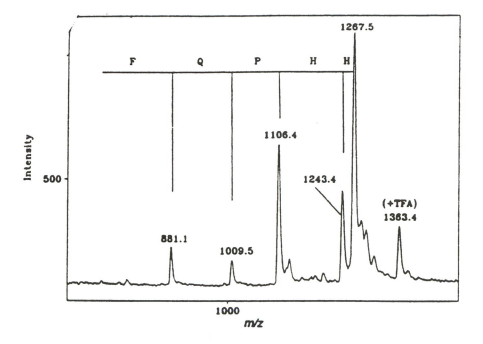

FIGURE 10.14   MALDI–TOF mass spectrum of a 5% portion of the peptide products isolated from a single resin bead. Reprinted with permission from reference 22.

FQPHHH. The mass differences between the capped species is straightforward. However, the difference in mass between the intact species at m/z 1267.5 and the first capped species at m/z 1243.4 must account for the mass of the CAP (113.0 Da). Thus, 1267.5 – 1243.4 + 113.0 = 137.1, which corresponds to a histidine residue.

Determination of Post-Translational Modifications

Expressed proteins are often modified by post-translational processing, which may include cleavages at specific residues; chemical modifications to the amino or carboxy terminus or to specific residues; formation of disulfide bonds; or the addition of phosphate, sulfate, fatty acyl groups, or carbohydrate. Chemical modifications may occur as well during sample isolation and preparation. When the sequence of the protein is known or can be inferred from the gene sequence, the type and location of specific modifications can often be deduced from simple mass differences between the expected and observed peaks in an enzymatic map. Thus, Table 10.4 shows a list of common modifications, their sites, and the mass differences that would be observed.

Often, the peptide fragments resulting from enzymatic digestion will have amino acid sequences that suggest more than one possible site. For example, the mass spectrum of an HPLC fraction from an Endo-Lys C digest of the 277 amino acid *assembly protein* (AP) of simian cytomegalovirus (sCMV) contained a peak at m/z 799.0 that was 80 mass units greater than that calculated (m/z 718.9) for the sequence: *SRSGALK*.[2] The addition of 80 mass units suggests that the protein is phosphorylated within this sequence which, however, contains two serine residues. In this case, the phosphorylation site may be determined simply by carrying out a

TABLE 10.4   Common post-translational modifications.

Name	Site	Modification	Change in mass, Δm
N-terminal acetylation	Terminal NH_2-	Replaced by CH_3CONH-	+42
N-terminal formylation	Terminal NH_2-	Replaced by $HCONH$-	+28
N-terminal myristylation	Terminal NH_2-	Replaced by $CH_3(CH_2)_{12}CONH$-	+210
N-terminal palmitoylation	Terminal NH_2-	Replaced by $CH_3(CH_2)_{14}CONH$-	+238
C-terminal amidation	Terminal -COOH	Replaced by $-CONH_2$	−1
Disulfide bonds	2 Cys -SH	Replaced by -S-S-	−2
Glycosylation (N-linked)	$\underline{N}$-X-S/T	See Table 10.7	
Glycosylation (O-linked)	$\underline{S/T}$	See Table 10.7	
Sulfation	-OH of Y	Replaced by $-OSO_3H$	+80
Phosphorylation	-OH of $Y/S/T$	Replaced by $-OPO_3H_2$	+80
N-methylation	$-NH_2$ of $K/R/H/Q$	Replaced by $-NHCH_3$	+14
O-methylesterification	-COOH of E/D	Replaced by $-COOCH_3$	+14
Carboxylation	$-NH_2$ of E/D	Replaced by $-NHOCH_3$	+30
Hydroxylation	$-NH_2$ of $P/K/D$	Replaced by -NHOH	+16

different or additional digest, such as trypsin, which would cleave at the arginine residue. Alternatively, sequencing using carboxy or aminopeptidase would also reveal the serine that is phosphorylated from the mass difference (87.1 + 80.0 = 167.1) between two of the peaks in the ladder. Because the Ser_3 residue is in fact phosphorylated, the mass spectrum of the carboxypeptidase digest would produce the following peaks:

    1   87.1 156.2 167.1   57.1   71.1 113.2 128.2   17   1
H — Ser — Arg — Ser — Gly — Ala — Leu — Lys — OH — H⁺ $M/Z = 799.0$
             |
            PO_3H

         1 156.2 167.1   57.1   71.1 113.2 128.2   17   1
        H — Arg — Ser — Gly — Ala — Leu — Lys — OH — H⁺ $M/Z = 711.9$
                 |
              PO_3H

             1 167.1   57.1   71.1 113.2 128.2   17   1
             H — Ser — Gly — Ala — Leu — Lys — OH — H⁺ $M/Z = 555.7$
                |
             PO_3H

                 1   57.1   71.1 113.2 128.2   17   1
                H— Gly — Ala — Leu — Lys — OH — H⁺ $M/Z = 388.6$

                    1   71.1 113.2 128.2   17   1
                  H— Ala — Leu — Lys — OH — H⁺ $M/Z = 331.5$

                       1 113.2 128.2   17   1
                    H— Leu — Lys — OH — H⁺ $M/Z = 260.4$

                           1 128.2   17   1
                      H— Lys — OH — H⁺ $M/Z = 147.2$

TABLE 10.5   Consensus sites for protein phosphorylation by kinases.

Enzyme	Consensus sequence
Protein kinase C	$(R/K_{1-3}, X_{2-0})$-S/T-$(X_{2-0}, R/K_{1-3})$
cAMP-dependent protein kinase	R-R-X-S/T
cGMP-dependent protein kinase	R/K_{2-3}-X-S/T
Casein kinase I	$S[P]$-X_{1-3}-S/T
Casein kinase II	S/T-$(D/E/S[P])_{1-3}$, X_{2-0}

X = unspecified amino acid; S[P] = phosphorylated serine

When post-translational modifications involve fatty acylation or glycosylation, it may also be of interest to determine the structure of the modification itself, as well as the specific site on the protein. In such cases (as described below) similar chemical and enzymatic strategies have been developed.

Phosphorylation Sites.   Protein phosphorylation occurs at serine and threonine residues, replacing the hydroxyl group with a phosphate. Phosphorylation is generally accomplished by enzymes, known as protein kinases (PK), which recognize specific amino acid sequences in the vicinity of the serine or threonine residue. Although these amino acid sequences are highly variable, it has been possible to establish *consensus sequences* for a number of protein kinases, which are shown in Table 10.5. For example, protein kinase C phosphorylates serine or threonine residues, flanked on both the C- and N-terminal sides by 0 to 2 amino acids that are followed by a combination of 1 to 3 arginine or lysine residues. Thus, in the example shown above, phosphorylation occurred within the (larger) amino acid sequence:

...RSRSGALKRRRER...

so that both of the serine residues were found within sequences that would be consensus sites for phosphorylation by protein kinase C.

In cases in which both a serine and threonine residue are present in a peptide fragment that is phosphorylated, these can be distinguished by reaction of the peptide with ethanethiol in 5N NaOH, which converts phosphoserine to *S*-ethylcysteine:

$$-NH\text{-}CH(CH_2PO_4)\text{-}CO\text{-} \rightarrow -NH\text{-}CH(CH_2\text{-}S\text{-}C_2H_5)\text{-}CO\text{-}$$

with the loss of 35 mass units. This reaction does not occur for phosphothreonine.

Carbohydrates and Glycoconjugates

One of the most common forms of protein modification is glycosylation, the addition of generally large carbohydrate structures to specific amino acid residues, which in some cases form the major portion of the molecular mass of these glycoproteins.

Oligosaccharides may be attached to peptides by amide or acyl linkages. The N-linked oligosaccharides are generally attached to asparagine residues found in the consensus sequence: Asn-X-Ser/Thr. These commonly involve an N-acetyl-glucosamine (GlcNAc) residue:

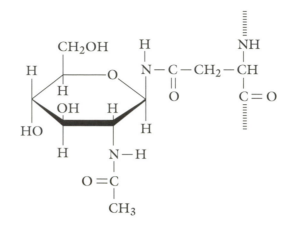

to which other sugars are attached. The O-linked oligosaccharides are attached to serine or threonine residues by an acyl linkage formed between two hydroxyl groups. An example is the linkage of an N-acetylgalactosamine (GalNAc) to serine:

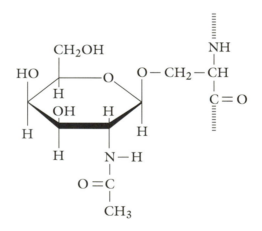

The remaining structure of an oligosaccharide is composed of additional sugar units, whose names, structures, and residue masses are listed in Table 10.6[23] Hexoses, such as glucose (Glc), galactose (Gal), and mannose (Man), have molecular weights of 180 Da. Formation of acyl linkages between hexoses and other sugar units is accompanied by the loss of water, so that the additional mass is 162 mass units. Similarly, amino sugars, such as glucosamine (GlcN) and galactosamine (GalN), result in the addition of 161 mass units, while N-acetylated sugars (GlcNAc, GalNAc) increase the mass of the oligosaccharide by 203 mass units. A number of sugars, such as glucuronic acid (GlcA) and N-acetylneuraminic acid (NANA, NeuAc, or *sialic acid*), include a carboxylic acid functional group. Glucuronic acid is formed by enzymatic oxidation of its uridine diphosphate (UDP) phosphoester in the presence of the

TABLE 10.6   Common sugars and their residue masses.

Monosaccharide	Abbreviation		Monoisotopic mass	Average mass
D-fructose	Fru		162.052	162.143
L-fucose	Fuc	F	146.058	146.14
D-galactose	Gal	GL	162.052	162.143
N-acetyl-D-galactosamine	GalNAc		203.079	203.19
D-glucose	Glc	G	162.052	162.143
D-glucuronic acid	GlcA		176.032	176.126
N-acetyl-D-glucosamine	GlcNAc		203.079	203.19
Hexose (nonspecific)	Hex		162.052	162.143
D-mannose	Man	M	162.052	162.143
N-acetyl-D-mannosamine	ManNac		203.079	203.19
N-acetylneuraminic acid	NANA	NeuAc SA	291.095	291.26
Sialic acid	SA		291.095	291.26

coenzyme nicotinamide-adenine dinucleotide (NAD). The resultant uridine diphosphate glucuronic acid (UDPGA) is the activated form of glucuronic acid that is the substrate for enzymatic attachment of GlcA to the oligosaccharide. UDPGA is also used in the enzyme-mediated metabolism of hydroxylated compounds occurring naturally in the body or those (for example, drugs or toxins) introduced exogenously, forming soluble glucuronide conjugates that can be excreted. Sialic acids (SA) are generally the terminal residues on branched complex carbohydrates. The presence of multiple anionic moieties is often troublesome for mass spectral analysis, particularly when accompanied by sodium counterions. Sialic acids can be removed enzymatically (using neuraminidases), or converted to their free acids by ion exchange with volatile ammonium salts, such as ammonium citrate.

N-Linked Oligosaccharides.   Processing of N-linked oligosaccharides is similar in yeast, plant, insect, and mammalian cells, and has been described in detail.[24,25] In this case, a fully assembled oligosaccharide linked through two phosphate groups to the lipid dolichol is transferred intact in the endoplasmic reticulum (ER) to an asparagine residue in the nascent protein. A series of enzymatic (exoglycosidase) reactions in the ER and Golgi trim the oligosaccharide to the *high mannose* structure:

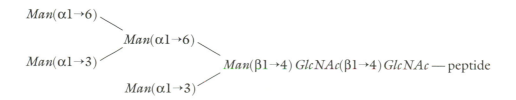

Using the activated GlcNAc-UDP as substrate, the enzyme N-acetylglucosamine-transferase I then attaches an N-acetylated glucosamine:

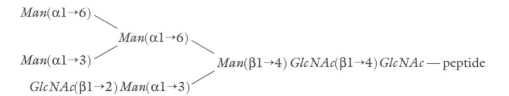

$Man(\alpha1\rightarrow6)$
$Man(\alpha1\rightarrow6)$
$Man(\alpha1\rightarrow3)$
$Man(\beta1\rightarrow4)\,GlcNAc(\beta1\rightarrow4)\,GlcNAc$ — peptide
$GlcNAc(\beta1\rightarrow2)\,Man(\alpha1\rightarrow3)$

Additional mannoses are then trimmed from the structure by mannosidases and, using activated GlcNAc-UDP, Gal-UDP, Fuc-GDP, and NeuAc-CMP (where GDP = guanosine diphosphate and CMP = cytidine monophosphate) as substrates, the enzymes N-acetylglucosaminetransferase II, galactosyltransferase, fucosyltransferase, and sialyltransferase assemble the final *complex carbohydrate* structures, such as the triantennary structure:

$NeuAc(\alpha1\rightarrow3)\,Gal(\beta1\rightarrow4)\,GlcNAc(\beta1\rightarrow2)\,Man(\alpha1\rightarrow6)$
$NeuAc(\alpha1\rightarrow3)\,Gal(\beta1\rightarrow4)\,GlcNAc(\beta1\rightarrow4)$
$Fuc(\alpha1\rightarrow6)$
$Man(\alpha1\rightarrow3)$
$Man(\beta1\rightarrow4)\,GlcNAc(\beta1\rightarrow4)\,GlcNAc$ — peptide
$NeuAc(\alpha1\rightarrow3)\,Gal(\beta1\rightarrow4)\,GlcNAc(\beta1\rightarrow2)$

While a number of N-linked complex carbohydrate structures are possible, they nearly always include several features of this structure. The linkage to asparagine is generally through two GlcNAc residues, followed by a branched three-mannose structure. The bi-, tri- and tetra-antennary carbohydrates are then formed by the addition of GlcNAc to the mannose residues in β1–3, β1–4, and β1–6 linkages. After addition of other sugar units, these branches are generally terminated by sialic acid. Fucosyl groups are generally found on the GlcNAc residue proximal to asparagine, but may be found as well on other GlcNAc residues. The complex carbohydrates attached to proteins from plants differ from those processed by mammalian cells. Plant-derived oligosaccharides do not contain sialic acids, but frequently contain xylose residues which are not found in mammalian glycoproteins.

O-Linked Oligosaccharides.   While the assembly of N-linked glycoproteins begins with the transfer of a fully assembled, lipid-linked oligosaccharide to an asparagine residue located within a consensus sequence, O-linked glycosylation begins with the attachment of a single GalNAc residue to a serine or threonine residue. The reaction is catalyzed by the enzyme UDP-GalNAc:polypeptide N-acetylgalactosaminetransferase with GalNAc-UDP as the substrate. Additional sugar residues are then assembled directly on the protein. In one example described by Goochee and co-workers,[25] the O-linked glycopeptide:

$NeuAc(\alpha2\rightarrow3)\,Gal(\beta1\rightarrow3)\,GalNAc(\alpha1)$ — Ser/Thr
$NeuAc(\alpha2\rightarrow6)$

is assembled from the substrates UDP-Gal and CMP-NeuAc.

TABLE 10.7   Glycosidases and their cleavage sites.

Endoglycosidases	Site
endo-β-N-acetylglucosaminidase F (Endo F)	GlcNAc(β1→4)GlcNAc
endo-β-N-acetylglucosaminidase H (Endo H)	GlcNAc(β1→4)GlcNAc
peptide-N-glycosidase F (PNGase F)	GlcNAc-Asn[a]

Exoglycosidases[b]	Site
β-galactosidase (jack bean)	Gal(β1→6>4≫3)GlcNAc
β-galactosidase (*S. pneumoniae*)	Gal(β1→4)GlcNAc,GalNAc
β-galactosidase (bovine testes)	Gal(β1→3>4≫3)GlcNAc,GalNAc
β-N-acetylglucosaminidase (*S. pneumoniae*)	GlcNAc(β1→2)Man, GlcNAc(β1→3,6)Gal
α-mannosidase (jack bean)	Man(α1→3,6)Man(β1→4)

[a] Converts Asn to Asp.

[b] From reference 30.

Mapping Carbohydrate Heterogeneity in Enzymatic Peptide Fragments. In contrast to modifications such as phosphorylation and sulfation, which result in a single change in mass, glycosylation generally involves the addition of carbohydrate species of varying length and composition at each site. When the protein sequence is known, the carbohydrate heterogeneity can be determined by carrying out a tryptic or other enzymatic digest of the protein, isolating the peptides carrying carbohydrate, and mapping those fragments by mass spectrometry. For example, the variant surface glycoprotein (VSG) of *Trypanosoma brucei* is a 59-kDa glycoprotein that has three glycosylation sites: a glycosylphosphatidylinositol anchor on the carboxy terminus of the protein;[26] an N-linked complex, biantennary oligosaccharide on Asn_{419}; and an N-linked high-mannose oligosaccharide on Asn_{432}. Initially, heterogeneity of the high-mannose site was determined by in vivo [³H]mannose-labeling of the glyco-peptide gp432:

$$Phe\text{-}Asn(GlcNAc\text{-}GlcNAc\text{-}Man_n)\text{-}Glu\text{-}Thr\text{-}Lys$$

obtained by digestion of the protein with trypsin–chymotrypsin, cleavage of the GlcNAc-GlcNAc linkage by Endo H (see Table 10.7), and size evaluation of the released carbohydrate by gel-filtration chromatography.[27] As shown in Figure 10.15, the need for radiolabeling and release of the free carbohydrate are both eliminated by the direct analysis of the intact gp432 glycopeptide using mass spectrometry. From this PDMS spectrum it can be determined that the gp432 glycopeptide carries 5 to 9 mannoses.[28]

Determination of Glycosylation Sites.   The enzyme endo-β-N-acetylglu-cosaminidase F (Endo F) also cleaves the glycosidic linkage between the two GlcNAc

species	monoisotopic mass	av mass
Phe-Asn(GlcNAc$_2$Man$_4$)-Glu-Thr-Lys	1692.68	1693.80
Phe-Asn(GlcNAc$_2$Man$_5$)-Glu-Thr-Lys	1854.73	1855.96
Phe-Asn(GlcNAc$_2$Man$_6$)-Glu-Thr-Lys	2016.78	2018.12
Phe-Asn(GlcNAc$_2$Man$_7$)-Glu-Thr-Lys	2178.84	2180.27
Phe-Asn(GlcNAc$_2$Man$_8$)-Glu-Thr-Lys	2340.89	2342.43
Phe-Asn(GlcNAc$_2$Man$_9$)-Glu-Thr-Lys	2502.94	2504.59

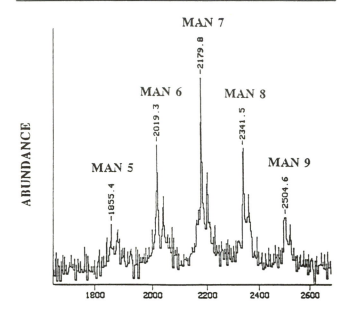

FIGURE 10.15   Positive ion plasma desorption TOF mass spectrum of the intact, nonradiolabeled gp432 glycopeptide from the 59-kDa VSG obtained from *T. brucei*. Reprinted with persmission from reference 28.

residues attached to asparagine, leaving a single GlcNAc residue attached to the peptide. For proteins whose amino acid sequences are known, mass mapping of an endoproteinase digest will reveal peaks for peptides containing a glycosylation site that indicate an increase in mass of 203 Da. The major limitations of Endo F are that it works effectively on high mannose, but not complex oligosaccharides, and it does not cleave O-linked oligosaccharides.[11]

The enzyme peptide-N-glycosidase F (PNGase F) cleaves the bond between GlcNAc and peptide, converting asparagine to aspartic acid. While this results in an increase in mass of only 1 Da, Gonzalez et al.[29] have carried out this reaction in the presence of 40% [^{18}O]-labeled H_2O and identified glycosylation sites by observing two peaks 1- and 3-Da larger than expected from the amino acid sequence. Because this approach works best when the isotopic species can be resolved, the protein should be digested to produce relatively small fragments in the vicinity of the expected site. Endo H, Endo F, and PNGase are all endogly-cosidases, and are listed in Table 10.7.

Enzymatic Methods for Determining Oligosaccharide Structures. Exoglycosidases can be used for determining carbohydrate structure in much the same way that exopeptidases are used to determine the amino acid sequence. However, while both types of enzymes work on terminal residues, there are some important differences between carbohydrate and peptide structures and the strategies that are used. With the exception of leucine–isoleucine and lysine–glutamine, amino acids have different masses, and are connected end-to-end in a single chain. Thus, carboxy or aminopeptidases function best when there is little discrimination for or against specific residues. In contrast, all hexoses have the same mass (as do GlcNAc and GalNAc), and are connected by different linkages to form highly branched structures. In this case, exopeptidases specific for a particular terminal residue and linkage (Table 10.7) can be used to probe the structure by mass spectrometry. For example, β-galactosidases are used to cleave terminal galactose from GlcNAc, while the source of the enzyme (jack bean, *S. pneumoniae*, bovine testes) provides specificity for β1→6, β1→4, and β1→3 linkages, respectively.[30] Other exoglycosidases include β-N-acetylglucosaminidases, α-mannosidases, and neuraminidases, which remove terminal GlcNAc, Man, and NeuAc residues, respectively.

The data shown in Figure 10.16 from Stahl et al.[30] illustrate the use of exoglycosidases to determine the structure and heterogeneity of a glycopeptide obtained by tryptic digestion of the purple acid phosphatase from the red kidney bean (KBPase). Figure 10.16a is the MALDI mass spectrum of the intact glycopeptide, revealing five peaks at m/z 4820 (I), 5023 (II), 5185 (III), 5224 (IV), and (5388) (V). The glycopeptides were resistant to cleavage of the oligosaccharide from the peptide using PNGase F, suggesting the presence of a fucosyl residue on the GlcNAc proximal to asparagine.[31] Proton-catalyzed defucosylation using 70% trifluoroacetic acid, resulted in shifting all five peaks by 146 Da, indicating that all species contained a fucosyl residue. Subsequent reaction with PNGase F resulted in removal of the oligosaccharides, giving the single peak at m/z 3651 shown in Figure 10.16b. This was consistent with the molecular weight calculated from the amino acid sequence (3649 Da), conversion of asparagine to aspartic acid, and protonation to form the MH+ ion.

Digestion of the intact glycopeptide mixture with β-galactosidase from jack bean (Figure 10.16c) and bovine testes reduced the glycopeptides to I, II, and IV (while that from S. pneumoniae did not), suggesting a Gal(β1→3)HexNAc in III and V. Subsequent digestion of this mixture with β-N-acetylglucosaminidase removed a single GlcNAc residue from structure II and two GlcNAc residues from structure IV, reducing all of the glycopeptides to structure I (Figure 10.16d). Digestion of this species with α-mannosidase at low enzyme concentration removed a single mannose residue, corresponding to cleavage of Man(α1→3)Man(β1→4), resulting in the peak at m/z 4659 in Figure 10.16e. Subsequent digestion with this same enzyme at high concentration resulted in loss of a second mannose, corresponding to cleavage of Man(α1→6)Man(β1→4). The peak at m/z 4497 in Figure 10.16f corresponds to a core glycan (consisting of 2 GlcNAc, 1 Man, 1 Fuc, and 1 Xyl residues). The final structures of the oligosaccharides I-V are given in Table 10.8, and include isomeric species that were determined by additional experiments.[30]

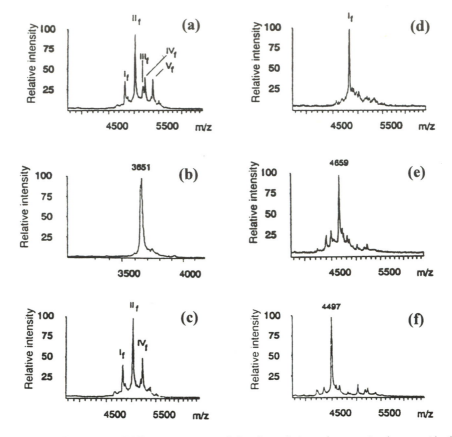

FIGURE 10.16   MALDI–TOF mass spectra of the degradation of a tryptic glycopeptide from KBPase. (a) Intact glycopeptide used in all subsequent digestions, (b) after digestion with PNGase F, (c) after digestion with jack bean β-galactosidase, (d) after digestion with jack bean β-galactosidase and β-N-acetylglucosaminidase from S. pneumoniae, (e) after digestion with jack bean β-galactosidase, β-N-acetylglucosaminidase from S. *pneumoniae* and jack bean α-mannosidase at low enzyme concentration, and (f) after digestion with jack bean β-galactosidase, β-N-acetylglucosaminidase from S. *pneumoniae* and jack bean α-mannosidase at high enzyme concentration. Reprinted with permission from reference 30.

Fragmentation Methods

Molecular weight measurements of peptides digested with exopeptidases form an important strategy for amino acid sequencing by time-of-flight mass spectrometry. However, as we have noted in Chapter 8, fragmentation resulting from post-source decay can be observed in reflectron instruments by stepping the reflectron voltage[32] or by using a nonlinear (curved-field) reflectron.[33] Thus, the post-source fragmentation of peptide ions forms an alternative approach to amino acid sequencing.

Fragmentation of Protonated Peptides.   The amino acid sequence-specific fragmentation of peptides is directed by the site of protonation. For peptides containing arginine and lysine, protonation occurs primarily on these basic residues. However, in order to anticipate the entire set of possible protonated fragments that may be

TABLE 10.8   Oligosaccharide structures for tryptic peptides from KBPase.[a]

Peak	Oligosaccharide structure	Composition %
$V_f(a)$	Galβ(1-3)GlcNAcβ(1-4)Manα(1-6) Manβ(1-4)GlcNAcβ(1-4)GlcNAc GlcNAcβ(1-4)Manα(1-3) Xylβ(1-2) Fucα(1-3)	12
$V_f(b)$	GlcNAcβ(1-4)Manα(1-6) Manβ(1-4)GlcNAcβ(1-4)GlcNAc Galβ(1-3)GlcNAcβ(1-4)Manα(1-3) Xylβ(1-2) Fucα(1-3)	3
IV_f	GlcNAcβ(1-4)Manα(1-6) Manβ(1-4)GlcNAcβ(1-4)GlcNAc GlcNAcβ(1-4)Manα(1-3) Xylβ(1-2) Fucα(1-3)	18
$III_f(a)$	Galβ(1-3)GlcNAcβ(1-4)Manα(1-6) Manβ(1-4)GlcNAcβ(1-4)GlcNAc Manα(1-3) Xylβ(1-2) Fucα(1-3)	8
$III_f(b)$	Manα(1-6) Manβ(1-4)GlcNAcβ(1-4)GlcNAc Galβ(1-3)GlcNAcβ(1-4)Manα(1-3) Xylβ(1-2) Fucα(1-3)	2
$II_f(a)$	GlcNAcβ(1-4)Manα(1-6) Manβ(1-4)GlcNAcβ(1-4)GlcNAc Manα(1-3) Xylβ(1-2) Fucα(1-3)	32

TABLE 10.8 (continued)   Oligosaccharide structures for tryptic peptides from KBPase.[a]

Peak	Oligosaccharide structure	Composition %

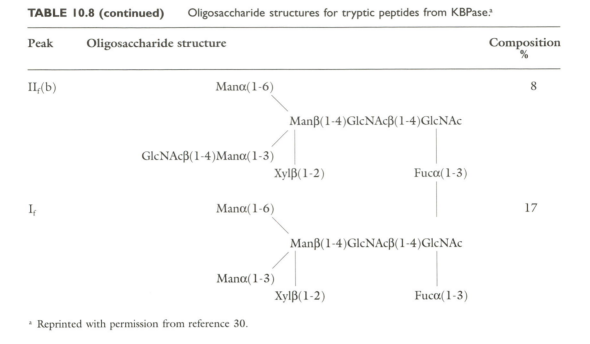

| II$_f$(b) | | 8 |
| I$_f$ | | 17 |

[a] Reprinted with permission from reference 30.

observed in the mass spectrum of a peptide, it is generally useful to consider that protonation can occur randomly on any one of the amide bonds, and is localized on the nitrogen atom:

$$\text{\tiny||||||||} NH - CH(R_1) - \underset{\underset{O}{||}}{C} - \overset{\oplus}{\underset{H}{N}}(H) - CH(R_2) - C\text{\tiny||||||||}$$

Because the protonated peptide is an even-electron ion, fragment ions will also generally be even-electron ions. Thus, if cleavage of the amide bond retains the two bonding electrons on nitrogen, **b** ions will be formed:

$$\text{\tiny||||||||} NH - CH(R_1) - C \equiv O \oplus$$

These ions can lose a neutral CO molecule to form **a** ions:

$$\text{\tiny||||||||} \overset{\oplus}{N}H = CH(R_1)$$

Both **a** and **b** ions involve retention of the positive charge on the amino-terminus, and provide a series of peaks that reveal the amino acid sequence from the carboxy-terminus. Alternatively, cleavage of the amide bond accompanied by H-transfer to the nitrogen atom results in formation of **y** ions:

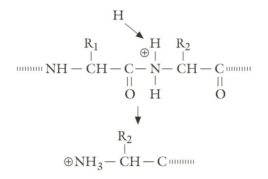

which can be used to determine the amino acid sequence from the amino-terminus. Of course, the **a** ions might have as easily been formed by direct cleavage of the C–C bond (rather than a two-step process). Thus, the entire set of peptide-chain cleavage ions can be described in a notation first suggested by Roepstorff et al.[34] and modified by Biemann[35] where the appropriate H-transfers (resulting in even-electron ions) are included:

$$\begin{array}{ccccccccc}
x_3 & y_3 & z_3 & x_2 & y_2 & z_2 & x_1 & y_1 & z_1 \\
\end{array}$$

$$\begin{array}{c}
R_1 \qquad\qquad R_2 \qquad\qquad R_3 \qquad\qquad R_4 \\
| \qquad\qquad\quad | \qquad\qquad\quad | \qquad\qquad\quad | \\
H_2N-CH-C-NH-CH-C-NH-CH-C-NH-CH-C-OH \\
\quad\;\; || \qquad\qquad || \qquad\qquad || \qquad\qquad || \\
\quad\;\; O \qquad\qquad\; O \qquad\qquad\; O \qquad\qquad O \\
\end{array}$$

$$\begin{array}{ccccccccc}
a_1 & b_1 & c_1 & a_2 & b_2 & c_2 & a_3 & b_3 & c \\
\end{array}$$

Sequence ions in Mass Spectra.   When the amino acid sequence is known, one can easily generate the set of possible **a**, **b**, **c**, **x**, **y**, and **z** ions that may be observed in a mass spectrum, and (indeed) there are now many computer programs available to do this. For example, one can calculate the set of **y** ions formed by protonated *DAEFR* from the scheme:

$$\begin{array}{cccccccc}
1 & 115.1 & 71.1 & 129.1 & 147.2 & 156.2 & 17 & 1 \\
H- & Asp- & Ala- & Glu- & Phe- & Arg- & OH & + H^+ \\
 & +H & +H & +H & +H & +H & & \\
\end{array}$$

by summing the masses to the right of each cleavage point and including the mass of a hydrogen atom. Thus, the largest **y** ion:

$$\begin{matrix} 115.1 & 71.1 & 129.1 & 147.2 & 156.2 & 17 & 1 \\ \text{Asp} & \text{Ala} & \text{Glu} & \text{Phe} & \text{Arg} & \text{OH} + \text{H}^+ \\ +1 \end{matrix}$$

115.1   71.1   129.1   147.2   156.2   17   1
Asp — Ala — Glu — Phe — Arg — OH + H⁺   M/Z = 637.7
+1

is the protonated molecular ion. The other **y** ions are then:

71.1   129.1   147.2   156.2   17   1
Ala — Glu — Phe — Arg — OH + H⁺   M/Z = 522.6
+1

129.1   147.2   156.2   17   1
Glu — Phe — Arg — OH + H⁺   M/Z = 451.5
+1

147.2   156.2   17   1
Phe — Arg — OH + H⁺   M/Z = 322.4
+1

156.2   17   1
Arg — OH + H⁺   M/Z = 175.2
+1

and are the same ions that would be generated by an aminopeptidase digestion of this peptide. In a similar manner, the **b** ions can be generated from the scheme:

1   115.1   71.1   129.1   147.2   156.2   17   1
H — Asp ┼ Ala ┼ Glu ┼ Phe ┼ Arg ┼ OH + H⁺

by summing the masses to the left of the arrow. (In this case there are no H-transfers). The largest **b** ion is:

1   115.1   71.1   129.1   147.2   156.2
H — Asp — Ala — Glu — Phe — Arg (+)   M/Z = 619.7

and can be easily recognized in the mass spectrum since it is 18 Da less than the protonated molecular ion; that is, it corresponds to the loss of a neutral water molecule, or MH⁺-H₂O. The remaining **b** ions are:

1   115.1   71.1   129.1   147.2
H — Asp — Ala — Glu — Phe (+)   M/Z = 463.5

1   115.1   71.1   129.1
H — Asp — Ala — Glu (+)   M/Z = 316.3

1   115.1   71.1
H — Asp — Ala (+)   M/Z = 187.2

1   115.1
H — Asp (+)   M/Z = 116.1

These ions are all 18 Da less than the ions that would be generated by a carboxypeptidase digest of this peptide. The **a** ions can then be generated by subtracting

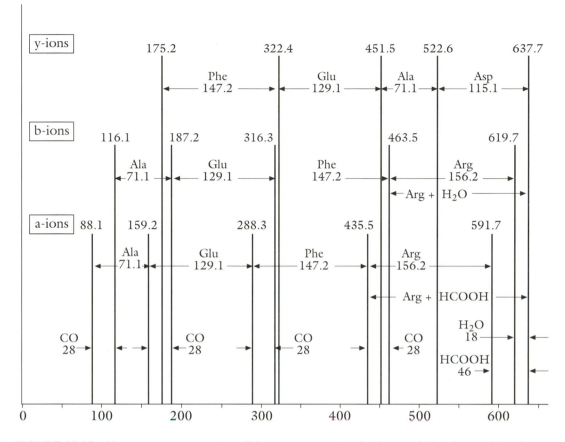

FIGURE 10.17 Histogram representation of the mass spectral peaks that would be observed for the **a**, **b**, and **y** ions of the peptide *DAEFR*.

28 mass units from each of the **b** ions, i.e., 591.7, 435.5, 288.3, 159.2, and 88.1. The largest **a** ion (m/z 591.7) could also be easily recognized in the mass spectrum, since its mass is 46 Da less than the protonated molecular ion; that is, it corresponds to the loss of a neutral formic acid molecule, or MH$^+$-HCOOH.

An idealized mass spectrum of the peptide *DAEFR* is shown in Figure 10.17. If this were a mass spectrum for a peptide whose amino acid sequence was unknown, it would be difficult indeed to determine which peaks correspond to **a**, **b**, or **y** ions. However, the series of **a** and **b** ions can often be recognized as pairs of peaks separated by 28 mass units. In addition; the protonated molecular ion MH$^+$ is always a **y** ion, while the peaks at m/z 619.7 (loss of water) and 591.7 (loss of HCOOH) can be clearly recognized as the largest **b** and **a** ions, respectively. Then, comparing the mass of the MH$^+$ peak with the peaks at m/z 522.6, 463.5, and 435.5 one obtains the mass differences:

$$637.7 - 522.6 = 115.1$$
$$637.7 - 463.5 = 173.6$$
$$637.7 - 435.5 = 202.2$$

where 115.1 corresponds to the mass of an aspartic acid residue, and other mass differences do not correspond to a specific residue. This begins the **y** ion series and indicates that the amino-terminal peptide is aspartic acid. Similarly, one begins the **b** series by comparing the MH$^+$-H$_2$O peak at m/z 619.7 with these same peaks to calculate the mass differences:

$$619.7 - 522.6 = 97.1$$
$$619.7 - 463.5 = 156.2$$
$$619.7 - 435.5 = 184.2$$

where only the mass difference of 156.2 Da corresponds to an amino acid residue, and indicates that the first carboxy-terminal peptide is arginine. When the MH$^+$-H$_2$O and MH$^+$-HCOOH ions are not observed, the carboxy-terminal peptide could be deduced from the mass differences from the protonated molecular ion:

$$637.7 - 463.5 - 18 = 156.2$$
$$637.7 - 435.5 - 46 = 156.2$$

Having established these residues, one then looks at the remaining mass differences in order to determine the amino acid sequence.

Peptides generally show a preference for one or more of the possible sequence ion series. Thus, the peptide *DAEFR*, which contains a carboxy-terminal arginine, can be expected to produce ions in which the charge is located on the C-terminus, that is **x**, **y**, or **z** ions. The production of fragment ions with the charge retained on the C-terminus is common for peptides digested with trypsin, which cleaves at arginine and lysine residues.

Internal and Side-Chain Cleavages. In addition to the formation of the six sequence-specific ions described above, a number of side-chain cleavages can also occur, forming the N-terminal **d** ions and the C-terminal **v** and **w** ions. An **a** ion in which the R substituent is an alkyl group can be written as:[36]

$$\overline{\underset{H-(NH-CHR-CO)_{n\text{-}1}-NH-\overset{\overset{\displaystyle CR_n\ ^aR_n^b}{\|}}{CH}}{H^+}}$$

where $R_n{}^a$ and $R_n{}^b$ are the beta substituents of the nth amino acid. The corresponding **d** ion has the structure:

$$\overline{\underset{H-(NH-CHR-CO)_{n\text{-}1}-NH-\overset{\overset{\displaystyle HCR_n^b}{\|}}{CH}}{H^+}}$$

Similarly, a **w** ion is derived from a **z** ion:

$$\underset{\overset{\displaystyle CR_n\ ^aR_n^b}{\underset{\displaystyle\|}{CH-CO-}}\overline{\underset{(NH-CHR-CO)_{n\text{-}1}-OH}{H^+}}}{}$$

and has the structure:

$$\overset{\overset{\displaystyle\text{HCR}_n\,^b}{\underset{\|}{}}}{\text{CH}-\text{CO}-}\underbrace{(\text{NH}-\text{CHR}-\text{CO})_{n\text{-}1}-\text{OH}}_{\text{H}^+}$$

The **d** and **w** ions are both useful for distinguishing leucine (R = -CH_2-$CH(CH_3)_2$) from isoleucine (R = -$CH(CH_3)CH_2CH_3$). The **v** ion is derived from the **y** ion:

$$\underbrace{\text{H}-(\text{NH}-\text{CHR}-\text{CO})_n-\text{OH}}_{\text{H}^+}$$

and has the structure:

$$\text{HN}=\text{CH}-\text{CO}-\underbrace{(\text{NH}-\text{CHR}-\text{CO})_{n\text{-}1}-\text{OH}}_{\text{H}^+}$$

In this case the entire side chain is lost.

A number of *non sequence-specific* ions also are observed in the mass spectra of peptides. These include internal acyl ions:

$$H_2N-\text{CHR}-\text{CO}-\text{NH}-\text{CHR}-\text{C}\equiv\text{O}\oplus$$

and internal immonium ions:

$$H_2N-\text{CHR}-\text{CO}-\overset{\oplus}{\text{NH}}=\text{CHR}$$

including single amino acid immonium ions:

$$\overset{\oplus}{H_2N}=\text{CHR}$$

For example, Figure 10.18 shows internal ions (*PQ, QI, PQI, PRP,* and *PRPQ*) in the low-mass portion of the mass spectrum.

Some Examples.   Earlier in this chapter, we discussed the isolation and identification of an antigen *AMAPRTLLL* bound to a novel murine Class IB MHC molecule Qa-1. In that example, the amino acid sequence was determined from the MALDI mass spectra of ladder peptides obtained from aminopeptidase *M,* carboxypeptidase *P,* and chymotrypsin digestions. As shown in Figure 10.19, the amino acid sequence can also be determined from the post-source decay MALDI mass spectrum of the intact peptide.[37] In this case, a curved-field reflectron was used, so that the entire mass spectrum was recorded without changing the reflectron voltage.

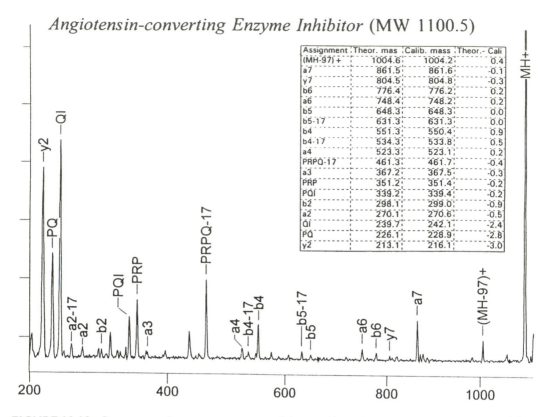

Angiotensin-converting Enzyme Inhibitor (MW 1100.5)

Assignment	Theor. mas	Calib. mass	Theor.- Cali
(MH-97)+	1004.6	1004.2	0.4
a7	861.5	861.6	-0.1
y7	804.5	804.8	-0.3
b6	776.4	776.2	0.2
a6	748.4	748.2	0.2
b5	648.3	648.3	0.0
b5-17	631.3	631.3	0.0
b4	551.3	550.4	0.9
b4-17	534.3	533.8	0.5
a4	523.3	523.1	0.2
PRPQ-17	461.3	461.7	-0.4
a3	367.2	367.5	-0.3
PRP	351.2	351.4	-0.2
PQI	339.2	339.4	-0.2
b2	298.1	299.0	-0.9
a2	270.1	270.6	-0.5
QI	239.7	242.1	-2.4
PQ	226.1	228.9	-2.8
y2	213.1	216.1	-3.0

FIGURE 10.18   Post-source decay mass spectrum of the peptide angiotensin-converting enzyme inhibitor (MW = 1100.5) obtained using a curved-field reflectron.

Fragment ions can also be observed by delayed-extraction, which increases the time that ions spend in the source. While delayed-extraction MALDI is currently being exploited for the amino acid sequencing of peptides, Figure 10.20a shows a delayed-extraction infrared laser desorption (IRLD) mass spectrum of cellotriose, obtained in 1990,[38] that provides considerable sequence-specific fragmentation. In this case, the fragment ions are derived from cationized species M+Na+ and M+K+. Figure 10.20b illustrates the nomenclature, similar to that used for peptide fragmentation, that has been introduced by Costello and Damon.[39] When the charge is retained on the nonreducing or distal end of the oligosaccharide, the fragment ions are designated as $^{r,s}A_n$, B_n and C_n, where n is the sugar residue numbered from the nonreducing end. Similarly, fragments in which the charge is retained on the reducing end of the oligosaccharide are designated as $^{r,s}X_n$, Y_n, and Z_n. Fragment ions $^{r,s}A_n$ and $^{r,s}X_n$ are formed by ring cleavages through the r and s bonds, and retain the proton or alkali ion of the molecular species. Fragment ions B_n and Y_n result from cleavage of the glycosidic (C-O) linkage, where B_n ions are oxonium ions:

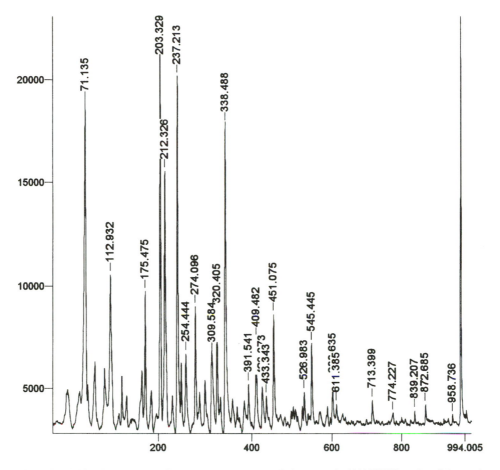

FIGURE 10.19 Post-source decay mass spectrum of the peptide *AMAPRTLLL* isolated from the Class IB MHC molecule Qa-1.

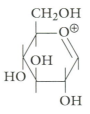

while Υ_n ions are protonated species that include a H-transfer:

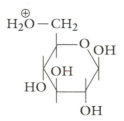

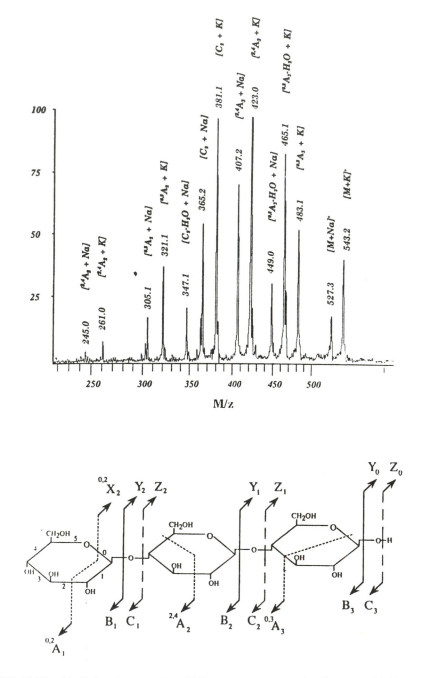

FIGURE 10.20 (a) Delayed extraction IRLD mass spectrum of cellotriose. (b) Structure of cellotriose Glc(β1→4)Glc(β1→4)Glc (MW = 504 Da), illustrating the fragment ion nomenclature according to reference 39. Reprinted with permission from reference 38.

C_n and Z_n ions result from cleavage of the glycosidic (O-C) bond. Z_n ions are rarely observed, while C_n ions are protonated and include an H-transfer:

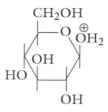

In many cases, the ions formed by retention of charge on the distal portion of the oligosaccharide can be distinguished from those in which the charge is retained on the reducing end by recognizing that B_n ions are always 18 mass units less that C_n ions, that is:

$$B_n + H_2O = C_n$$

analogous to the 28-mass-unit differences between *a* and *b* ions in peptide mass spectra. In addition, ring fragmentation can often be used to determine linkage positions.[38] Table 10.9 summarizes the cleavages observed in a series of 1→6, 1→4, and 1→3 linked sugars. The mechanisms for forming ions by two-bond ring cleavages have been described by Spengler, et al.[38] and Leary et al.[40]

An Integrated Approach.   Fragmentation methods do not obviate the need for the enzymatic and molecular weight strategies that have been devised to determine the structures of proteins, glycoproteins, and carbohydrates. While post-source decay methods provide an important opportunity for mass selection in mixtures (which delayed extraction does not), *de novo* sequencing is problematic, when the types of fragmentation cannot be easily determined and mass spectra contain an abundance of non sequence-specific internal ions. In such cases, the determination of one or more terminal residues from carboxy or aminopeptidase digestions can be a considerable aid in interpreting PSD mass spectra of peptides. In the case of carbohydrates, enzymatic approaches take on an additional importance, since fragmentation cannot distinguish isomeric species. Although considerable progress has been made (using two-bond cleavages) to distinguish linkages, fragmentation cannot distinguish between different hexoses. Thus, an integrated approach, that uses both enzymatic and fragmentation strategies, remains the best approach for the mass-spectral analysis of proteins, peptides, and glycoconjugates.

TABLE 10.9 Specificity of ring cleavages.

Substance	Mass losses	$^{0,2}A_n$	$^{0,4}A_n$	$^{2,4}A_n$	C_{n-1}	etc.	$^{0,3}A_n$	M – H$_2$O	$^{0,2}A_n$ – H$_2$O	C_{n-1} – H$_2$O
Cellobiose = Glc[β1→4]Glc	-60, -120, (-162),[a] (-18), (-60-18)	×		×	(×)			(×)	(×)	
Cellotriose = GLC[β1→4]Glc-[β1→4]Glc	-60, -120, -162, etc. (-18), -60-18, -162-18	×		×	×	×		(×)	×	×
Maltose = Glc[α1→4]Glc	-60, -120	×		×						
Maltotriose = Glc[α1→4]Glc-[α1→4]Glc	-60, -120, -162, etc.	×		×	×	×				
Gal[α1→4]Gal	-60, -120, (-162)	×		×	(×)					
Gal[β1→4]Gal-[β1→4]Glc	-60, -120, -162, etc. -18, -60-18, -162-18	×		×	×	×		×	×	×
Gentiobiose = Glc[β1→6]Glc	-60, -120, (-162) (-90), -18	×	×		(×)		(×)	×		
Isomaltose = Glc[α1→6]Glc	-60, -120, (-162), (-90)	×	×		(×)		(×)			
Isomaltotriose = Glc[α1→6]Glc-[α1→6]Glc	-60, -120, -90, -162, etc., -18	×	×		×	×	×	(×)		
Nigerose = Glc[α1→3]Glc	(-90), -162, (-162-18)	(×)			×		(×)			(×)
Mannobiose = Man[α1→3]Man	(-90), -162, (-60), (-120), (-162-18)			(×)	×		(×)			(×)
Trehalose = Glc[α1→1]Glc	(-162)				(×)					
Sucrose = Glc[α1→2]Fru	-162, -18	×			×			(×)		
Palatinose = Glc[α1→6]Fru	-90									
Raffinose = Gal[α1→6]Glc-[α1→2]Fru	-162, etc.				×	(×)				
γ-cyclodextrin = c-(Glc[α1→4])$_6$	-102, -102-60, -102-120, -102-162, etc.	×		(×)	×	×				

[a] Parentheses indicate that ions of low intensity were observed.

References

1. Larsen, B.S.; Yergey, J.A.; Cotter, R.J., *Biomed. Mass Spectrom.* **12** (1985) 586–587.
2. Woods, A.S.; Gibson, W.; Cotter, R.J., *Protein Processing in Herpesviruses*, in *Time-of-Flight Mass Spectrometry*, Cotter, R.J., Ed.; ACS Symposium Series, Washington, DC, (1994) pp 194–210.
3. Yergey, J.; Heller, D.; Hansen, G.; Cotter, R.J.; Fenselau, C., *Anal. Chem.* **55** (1983) 353–356.
4. Roher, A.E.; Lowenson, J.D.; Clarke, S.; Woods, A.S.; Cotter, R.J.; Gowing, E.; Ball, M.J., *Proc. Natl. Acad. Sci.* **90** (1993) 10836–10840.
5. Imperial, J.S.; Chaudhary, T.; Beavis, R.C.; Chait, B.T.; Hunsperger, J.P.; Olivera, B.M.; Adams, M.E.; Hillyard, D.R., *J. Biol. Chem.* **267** (1992) 20701–20705.
6. Terwilliger, T.C.; Koshland, D.E., *J. Biol. Chem.* **259** (1984) 7719.
7. Naylor, S.; Findeis, A.F.; Gibson, B.W.; Williams, D.H., *J. Am. Chem. Soc.* **108** (1986) 6339.
8. Allen, G., *Sequencing of Proteins and Peptides*, Elsevier, Amsterdam (1990) pp. 73–104.
9. Chevrier, M.R.; Cotter, R.J., *Rapid Commun. Mass Spectrom.* **5** (1991) 611–617.
10. Zhao, Y.; Chait, B.T., *Anal. Chem.* **66** (1994) 3723–3726.
11. Flannery, A.V.; Beynon, R.J.; Bond, J.S., in *Proteolytic Enzymes: A Practical Approach*, Beynon, R.J.; Bond, J.S., Eds.; IRL Press, Oxford (1990) pp. 148–149.
12. Patterson, D.H.; Tarr, G.E.; Regnier, F.E.; Martin, S.A., *Anal. Chem.* **67** (1995) 3971–3978.
13. Woods, A.S.; Cotter, R.J. (unpublished results).
14. Aldrich, C.J.; DeCloux, A.; Woods, A.S.; Cotter, R.J.; Soloski, M.J.; Foreman, J., *Cell* **79** (1994) 649–658.
15. Strupat, K.; Karas, M.; Hillenkamp, F.; Eckerskorn, C.; Lottspeich, F., *Anal. Chem.* **66** (1994) 464–470.
16. Henzel, W.J.; Billeci, T.M.; Stults, J.T.; Wong, S.C.; Grimley, C.; Watanabe, C., *Proc. Natl. Acad. Sci. USA* **90** (1993) 5011–5015.
17. Vestling, M.; Fenselau, C., *Anal. Chem.* **66** (1994) 471–477.
18. Chait, B.T.; Chaudhary, T.; Field, F.H., *Methods in Protein Sequence Analysis 1986*, Walsh, K.A. Ed., Humana Press, Clifton, NJ (1987) pp. 483–492.
19. Wang, R.; Cotter, R.J.; Meschia, J.F.; Sisodia, S.S., *Techniques in Protein Chemistry III*, Academic Press, San Diego, (1992) pp. 505–513.
20. Chait, B.T.; Wang, R.; Beavis, R.C.; Kent, S., *Science* **262** (1993) 89.
21. Vorm, O.; Roepstorff, P., *Biol. Mass Spectrom.* **23** (1994) 734–740.
22. Youngquist, R.S.; Fuentes, G.R.; Lacey, M.P.; Keough, T., *Rapid Commun. Mass Spectrom.* **8** (1994) 77–81.
23. Harvey, D.J., *American Laboratory* (December 1994) 22–28.
24. Kornfield, R.; Kornfield, S., *Ann. Rev. Biochem.* **54** (1985) 631–664.
25. Goochee, C.F.; Gramer, M.J.; Andersen, D.C.; Bahr, J.B.; Rasmussen, J.R., *Biotechnology* **9** (1991) 1347–1355.
26. Ferguson, M.A.; Homans, S.W.; Dwek, R.A.; Rademacher, T.W., *Science* **239** (1988) 743–759.
27. Bangs, J.D.; Doering, T.L.; Englund, P.T.; Hart, G.W., *J. Biol. Chem.* **263** (1988) 17697–17705.
28. Bean, M.F.; Bangs, J.D.; Doering, T.L.; Englund, P.T.; Hart, G.W.; Fenselau, C.; Cotter, R.J., *Anal. Chem.* **61** (1989) 2686–2688.
29. Gonzalez, J.; Takao, T.; Hori, H.; Besada, V.; Rodriguez, R.; Padron, G.; Shimonishi, Y., *Anal. Biochem.* **205** (1992) 151–158.
30. Stahl, B.; Klabunde, T.; Witzel, H.; Krebs, B.; Steup, M.; Karas, M.; Hillenkamp, F., *Eur. J. Biochem.* **220** (1994) 321–330.
31. Tretter, V.; Altmann, F.; März, L., *Eur. J. Biochem.* **199** (1991) 647–652.

32. Kaufmann, R.; Kirsch, D.; Spengler, B., *Int. J. Mass Spectrom. Ion Processes* **131** (1994) 355–385.

33. Cornish, T.J.; Cotter, R.J., *Rapid Commun. Mass Spectrom.* **8** (1994) 781–785.

34. Roepstorff, P.; Fohlman, J., *Biomed. Mass Spectrom.* **11** (1984) 601.

35. Biemann, K., *Biomed. Mass Spectrom.* **16** (1988) 99.

36. Biemann, K., in *Methods in Enzymology 193: Mass Spectrometry*, McCloskey, J.A., Ed.; Academic Press, San Diego (1990) pp. 886–887.

37. Woods, A.S.; DeCloux, A.; Cotter, R.J.; Soloski, M.J. (unpublished results).

38. Spengler, B.; Dolce, J.W.; Cotter, R.J., *Anal. Chem.* **62** (1990) 1731–1737.

39. Domon, B.; Costello, C., *Glycoconj. J.* **5** (1988) 397–409.

40. Hofmeister, G.E.; Zhou, Z.; Leary, J.A., *J. Am. Chem. Soc.* **113** (1991) 5964–5970.

Oligonucleotides and the Human Genome

In a series of three papers in 1982, McNeal et al.[1-3] reported the positive- and negative-ion plasma desorption mass spectra of a series of fully protected synthetic oligonucleotides. Perhaps the first example of the potential for DNA sequencing by time-of-flight mass spectrometry, these spectra revealed abundant fragmentation, and an example of their negative-ion plasma desorption mass spectrum of the 6-mer CAACCA is shown in Figure 11.1. The trichloroethyl ester ($-OCH_3CCl_3$) derivatives used in synthesis contributed to their success in analyzing these oligomers, since this eliminated the multiple negative charges on the phosphate groups. The plasma desorption mass spectrometry analysis of unprotected oligonucleotides proved to be far more difficult. At least one example was provided by Vigny et al.[4] in 1987. Figure 11.2 shows their PDMS mass spectra of two trinucleotides GGT and TTT which are distinguished by their fragmentation. In 1990, Spengler et al.[5] utilized the new MALDI technique for the analysis of deoxyribonucleotide oligomers up to the 7-mer. The time-of-flight MALDI mass spectrum of the tetramer TTGG is shown in Figure 11.3, and was obtained by applying delayed-extraction *pushout* and *drawout* pulses to the backing plate and extraction grids, respectively. In this case, only the molecular ion is observed. This has proven to be typical of MALDI mass spectra of oligonucleotides. Thus, most of the approaches to sequencing DNA oligomers have focused on *ladder* methods.

Sequencing the Human Genome

The oligonucelotides forming the doubly stranded helical structure of DNA are composed of four bases: adenine (A), cytosine (C), thymine (T), and guanine (G), attached to deoxyribose sugar units that are connected by phosphate groups. Figure 11.4 shows the structure of an oligonucleotide containing each of these bases, along with the abbreviated structural notation that will be used in this chapter. The oligonucleotide strands in DNA are complementary in their base sequences, connected by hydrogen bonds between adenine and thymine (A-T) and cytosine and

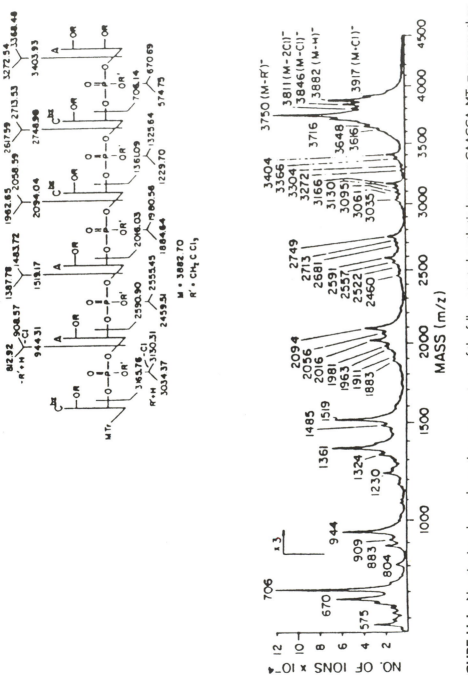

FIGURE 11.1   Negative ion plasma desorption mass spectrum of the fully protected synthetic hexamer CAACCA. MTr = monomethoxytrityl, C^{bx} = benzoylcytidine, and R = TBDMS. (Reprinted with permission from reference 2).

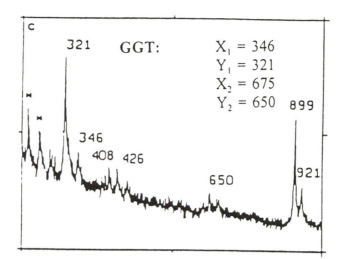

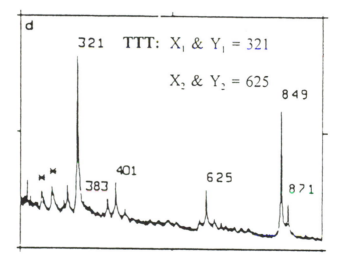

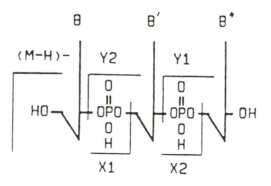

FIGURE 11.2 Plasma desorption mass spectra of the unprotected trinucleotides (a) GGT and (b) TTT. (c) Notation used to describe fragmentation. (Reprinted with permission from reference 4).

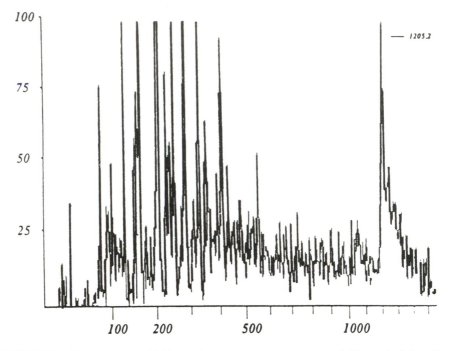

FIGURE 11.3 Matrix-assisted UV laser desorption mass spectrum of 40 pmol of the oligode-oxyribonucleotide TTGG (MW = 1204 Da). (Reprinted with permission from reference 5).

guanine (C-G). Currently, the most widely used method for sequencing DNA is based upon a method developed by Sanger et al.[6] in which four dideoxy-terminated DNA ladders are synthesized by the polymerase chain reaction using the complementary strand as a template. For example, synthesis of the *T*-ladder is carried out using the deoxyribose nucleosides (dA, dC, dT, and dG) and the dideoxy terminator (ddT) as reagents. As shown in Figure 11.5, the resultant ddT-terminated oligonucleotides (which include a primer) are separated by gel electrophoresis, and compared with the ddA-, ddC-, and ddG-terminated ladders to determine the sequence. Interestingly, while gel electro-phoresis separates the oligomers roughly according to mass (size), it is the order of their appearance in each of the four lanes that is used to determine the sequence.

The human genome consists of approximately 3.3×10^9 base-pairs. Electro-phoretic methods for separating oligonucleotide ladders currently have a resolution of about 600 bases. Given the time required for carrying out the separations, it has been estimated that a typical laboratory might sequence about 30,000 bases/year, so that the efforts of 1000 such laboratories would require 110 years to sequence the entire human genome using current technologies. This plus the interest in comparative sequencing of many individuals, viruses, bacteria, and other species underscores the need for new, more efficient technologies. Because of its ability to measure the masses of large biopolymers, MALDI time-of-flight mass spectrometry has received considerable attention. At the same time, the rapid success of this technique for analyzing peptides and proteins has not as easily been attained for the analysis of oligonucleotides in the same mass range. In contrast to proteins, the absence of a defined tertiary structure, the presence of multiple anionic phosphate

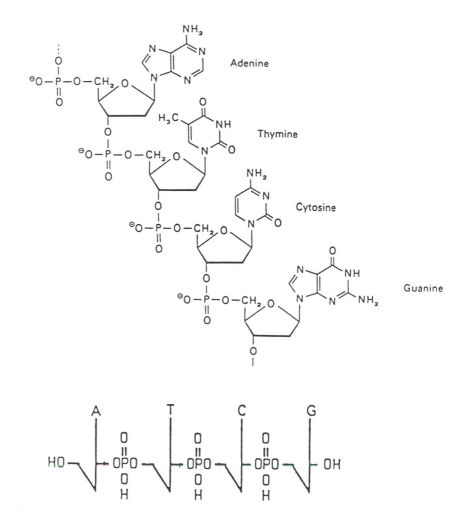

FIGURE 11.4   (a) Structure of the oligonucleotide d(ACTG). (b) Abbreviated notation.

groups, the tendency to form ion-pairs with sodium, and non-sequence-specific fragmentation adversely affect both the mass range (oligomer size) that can be analyzed and the mass resolution.

Most of the approaches that have been proposed for DNA sequencing using MALDI–TOF mass spectrometry have been ladder approaches, recording the molecular ions of a mixture of DNA (or RNA) oligomers. If one considers the masses of the bases (X) and their respective deoxyribonucleotides (dXp) and ribonucleotides (Xp):

adenine	A	135.1	dAp	313.2	Ap	329.2
cytosine	C	111.1	dCp	289.2	Cp	305.2
thymine	T	126.1	dTp	304.2		
guanine	G	151.1	dGp	329.2	Gp	345.2
uracil	U	112.1			Up	306.2

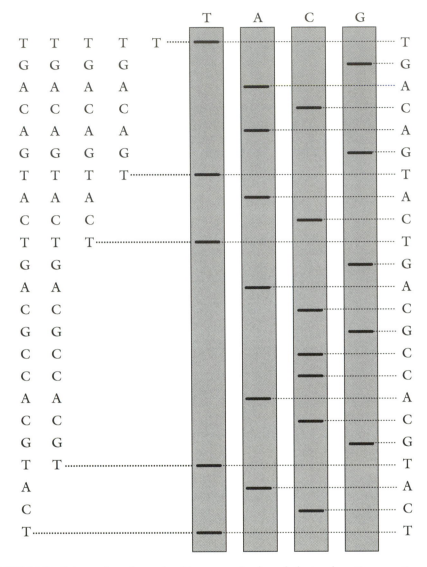

FIGURE 11.5 Scheme for oligonucleotide sequencing by gel-electrophoretic separation of four dideoxy-terminated Sanger ladders.

the ability to distinguish A from T in a 30-mer DNA would require a mass resolution greater than 1 part in 1000 at a mass of 9000. Similarly, the ability to distinguish U from C in a 30-mer RNA would require a mass resolution greater than 1 part in 9000. At this mass, such resolutions are difficult to obtain on TOF instruments (even for proteins!). However, such stringent requirements on resolution (and mass accuracy) are reduced if the mass spectra are recorded for four separate dideoxy-terminated ladders, in which the identity of the terminal oligonucleotide is determined from its presence in a particular ladder. In the scheme shown in Figure 11.6, the mass resolution required to separate the 15- and 16-mers in the C-ladder spectrum (assuming that the primer is a 25-mer) would be 1 part in 40. Thus, the common

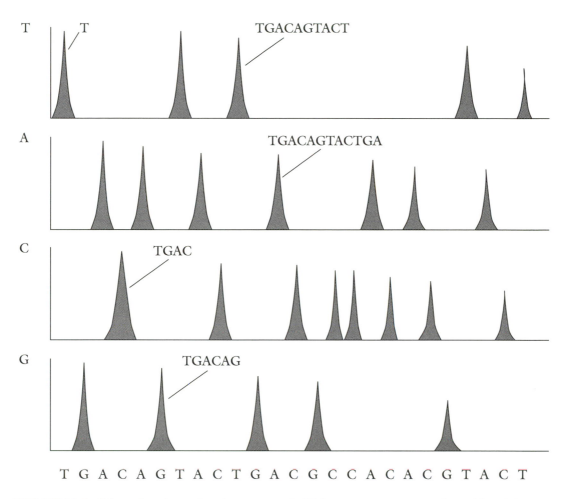

FIGURE 11.6 Scheme for oligonucleotide sequencing by TOF mass spectrometric analysis of four dideoxy-terminated Sanger ladders.

objective has been to utilize Sanger methods in combination with MALDI–TOF mass spectrometry, in which the mass spectral measurement reduces the time required for electrophoretic separations.

In this case, the major challenge has been to improve the mass range and resolution for the analysis of oligomers by MALDI–TOF to enable this technique to be competitive with existing electrophoretic methods that can resolve oligomers up to the 600-mer and beyond. These efforts are the first consideration of this chapter, which deals with the effects of matrices, laser wavelengths, the reduction of sodium counterions, and instrumentation on mass range and resolution. We then consider the status of Sanger ladder-sequencing methods to date, the generation of ladders by controlled hydrolysis and exonucleases, and fragmentation methods. Finally, we consider methods in which oligomers are hybridized to probes attached covalently to beads or surfaces, and mass spectral strategies that fit well within the existing and probable future capabilities of MALDI–TOF mass spectrometry.

Improving the Mass Range and Resolution of MALDI–TOF Mass Spectrometry of Oligonucleotides

A major reason for the success in desorbing intact proteins by MALDI is the fact that the UV radiation is absorbed primarily by the matrix, and not the analyte. In contrast, oligonucleotides are composed of chromophoric nucleotide bases that absorb in the same spectral region as the matrix, leading to extensive fragmentation and reduction in molecular-ion intensities. Compared with proteins, the desorption efficiency of oligonucleotides decreases much more rapidly with oligomer size. While proteins have a generally well-defined tertiary structure (generally globular) and are hydrophobic, the negatively charged phosphate groups in oligonucleotides produce extended structures that form ionic interactions with the matrix. The formation of multiple adducts with counterions (primarily sodium) reduces the mass measurement accuracy when these are unresolved, as does the loss of nucleotide bases by fragmentation. Unlike peptides, which fragment along the peptide chain and provide information on the amino acid sequence, the facile loss of one or more nucleotide bases, attached as side groups to the phosphodeoxyribose backbone, gives no information about their location. Additionally, a number of investigators have observed that poly(dT) oligomers desorb more efficiently than poly(dA, dC, or dG) or mixed-base oligomers.[7,8]

Matrices and Wavelengths.   Considerable effort has been expended in exploring different matrix and laser wavelength combinations that would improve desorption efficiency, the size range of oligomers that can be desorbed, and mass resolution. Figure 11.7 shows an early MALDI–TOF mass spectrum of a 24-mer DNA probe d(CATGTCAAAATTACAGACTTCGGG) obtained using a nitrogen laser (337 nm) and gentisic acid as the matrix.[9] Using a 45-kV prompt-extraction TOF mass spectrometer, Tang et al.[10] compared the desorption of a mixture of mixed-base oligomers (from the 3-mer to 34-mer) using the second, third, and fourth harmonics (532, 355, and 266 nm, respectively) of a Nd:YAG laser and matrices appropriate to these wavelength regions. As shown in Figure 11.8, all oligomers were observed at 355 nm, while neither the 20-mer or 34-mer was observed at 532 nm (where it is also possible that rhodamine is not as effective as a sample matrix). For the 266-nm (frequency-quadrupled) wavelength commonly used in MALDI, one of the most successful matrices has been a mixture of picolinic and 3-hydroxypicolinic acids. Figures 11.9a and 9b show the negative-ion MALDI–TOF mass spectra of two doubly stranded DNA oligomers having 246 and 500 base-pairs, respectively, obtained by Tang et al.[11] While these represent some of the largest oligomers desorbed and mass analyzed to date, the mass resolution is poor, in part because these doubly stranded DNA are desorbed as two unresolved single strands having the same number of bases but different masses. Cleavage of the 246-base-pair DNA with different restriction enzymes produces the restriction mass maps shown in Figure 11.10.[12]

Hillenkamp et al.[13] have explored the use of an Er:YAG (2.94 μm) infrared laser for the desorption of DNA oligomers and compared their results with those obtained at other wavelengths. Figure 11.11a shows the negative-ion mass spectrum of a mixture of poly-d(T)$_{19-24}$ oligomers using a UV nitrogen laser (337 nm). While the major species desorbed are (M-H)$^-$ ions, the mass resolution is compromised by the unresolved contributions from (M-2H+Na)$^-$ ions. Desorption using the Er:YAG laser (Figure 11.11b) reveals primarily (M-2H+Na)$^-$ ions, while the addition of cation-

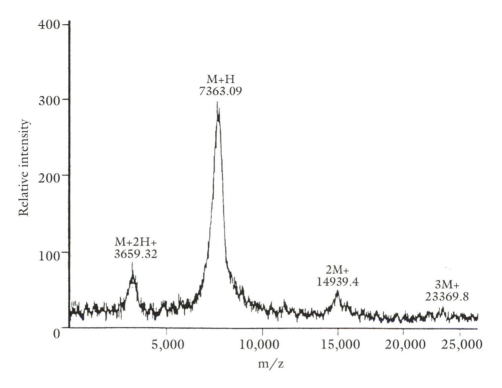

FIGURE 11.7   MALDI–TOF mass spectrum of d(CATGTCAAAATTACAGACTTCGGG) desorbed from gentisic acid using a pulsed nitrogen laser at 337 nm.

exchange beads (to remove Na$^+$ ions, Figure 11.11c) produces the most well-resolved mass spectrum. While these results would appear to give far more support to the need for desalting than for the use of laser wavelengths in the IR region, we have noted that poly-T oligomers are particularly easy to desorb. In another example reported by this group, they compared the UV and IR laser desorption mass spectra of d(C)$_{17}$. UV desorption gave very poor results (Figure 11.12a), while IR desorption revealed a molecular-ion peak broadened by species containing multiple sodium and potassium cations and losses of one or more cytosine bases (Figure 11.12b). Addition of cation-exchange beads reduced the contributions from species containing alkali ions (Figure 11.12c) and improved the resolution, revealing very clearly the nonspecific losses of 1–3 cytosine bases. Thus, these results illustrate very clearly that the major barriers to achieving high desorption efficiency and mass resolution are the presence of alkali counterions and the loss of nucleotide bases.

Desalting.   The removal of alkali counterions has proven to be important for all wavelengths and matrices. As described above, this can be accomplished by the addition of cation-exchange beads. Alternatively, sodium and potassium ions can be removed by the simple addition of volatile ammonium salts to the analyte solution. Figure 11.13 compares the negative ion mass spectra of the mixed-base oligomer 5′-d(AGCTAGCT)–3′ obtained without and with the addition of diammonium hydrogen citrate.[14]

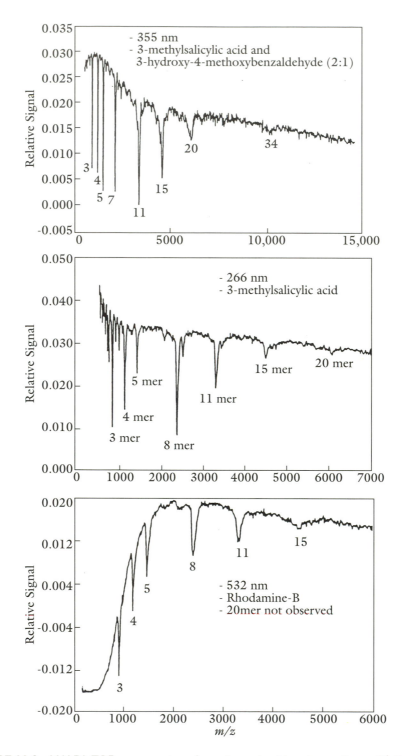

FIGURE 11.8 MALDI–TOF mass spectra of an oligonucleotide mixture: 3-mer (5'-AGT-3'), 4-mer (5'-AGTC-3'), 5-mer (5'-AGTCC-3'), 7-mer (5'-AGTCCTG-3'), 8-mer (5'-AGTCCTGA-3'), 11-mer (5'-AGTCCTGAAGT-3'), 15-mer (5'-AGTCCTGAAGTCCTG-3'), 20-mer (5'-AGTCCT-GAAGTCCTGAAGTC-3'), and 34-mer (5'-AGTCCTGAAGTCCTGAAGTCAGTCCT-GAAGTCCT-3'), using different wavelengths and matrices. (a) 355 nm and a 2:1 mixture of 3-methylsalicylic acid and 3-hydroxy-4-methoxybenzaldehyde, (b) 266 nm and 3-methylsalicylic acid, and (c) 532 nm and Rhodamine B. (Reprinted with permission from reference 10).

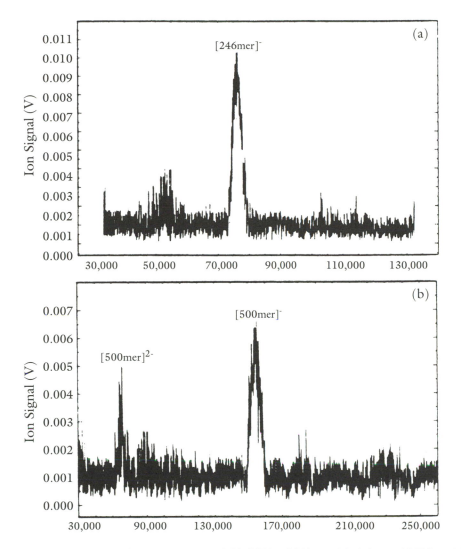

FIGURE 11.9   MALDI–TOF mass spectra of (a) 246-bp DNA amplified from pLB129 and (b) 500-bp DNA amplified from bacteriophage lambda genome. The laser wavelength was 266 nm and the matrix HPA/3-HPA. (Reprinted with permission from reference 11).

Nonspecific Base Losses and Mass Resolution.   Becker et al.[15] have compared the linear and reflectron mass spectra of oligonucleotides of different size using both low-voltage (3 kV) and high-voltage (12 kV) extraction. In the linear mode, high-voltage extraction produces far better mass resolution, since both the initial kinetic energy spread and fragmentation in the accelerating field will compromise resolution. Their comparison of high-voltage extraction linear and reflectron spectra for a mixed-base oligomer (shown in Figure 11.14) demonstrates considerably better mass resolution for reflectron mass spectra. In addition, the reflectron mass spectrum (Figure 11.14b and insert) resolves the nonspecific base losses that contribute to loss of mass resolution in linear TOF mass spectra.

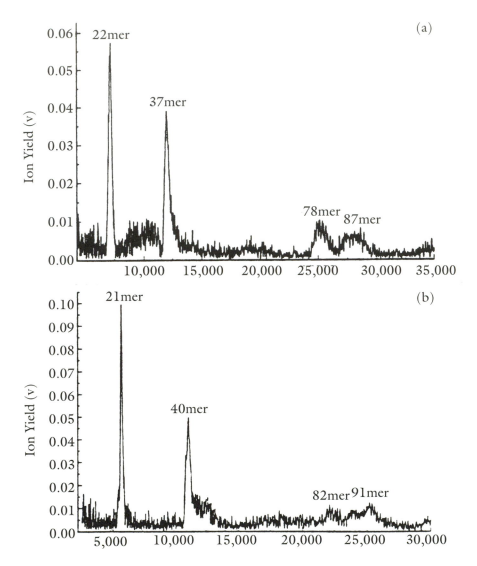

FIGURE 11.10 MALDI–TOF mass spectra of the 246-bp fragment amplified from pLB132 following cleavage with (a) restriction enzymes **StyI (C ↓ CATGG)** and **HinfI (G ↓ ANTC)** and (b) restriction enzyme **DdeI (C ↓ TNAG)**. (Reprinted with permission from reference 12).

Frozen Ice Matrices. Williams et al.[16] have suggested the possibility of desorbing oligonucleotides from frozen aqueous solutions corresponding to the normal biological matrices in which they are found. Using an argon-ion laser (589 nm) they obtained a MALDI mass spectrum of the doubly stranded DNA oligonucleotide:

5′-(AGCCCGCCTAATGAGCGGGCTTTTTTTT)–3′
3′-(TCGGGCGGATTACTCGCCCGAAAAAAAA)–5′

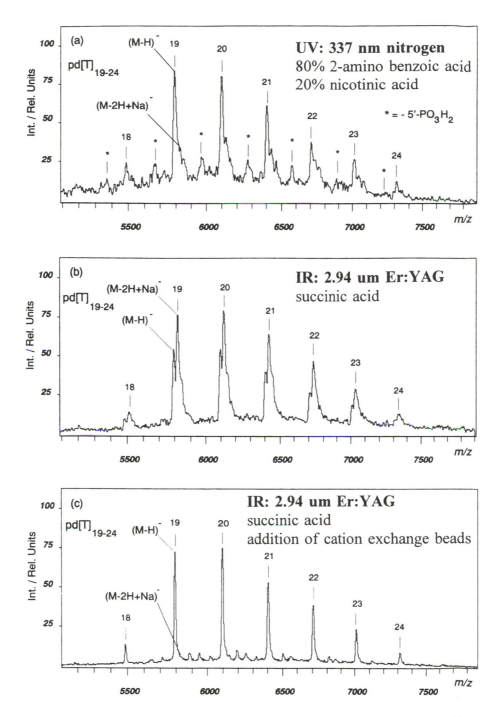

FIGURE 11.11   Comparison of the MALDI–TOF mass spectra of d(T)$_{19-24}$ using (a) a UV laser, (b) an IR laser, and (c) and IR laser following the addition of cation-exchange beads. (Reprinted with permission from reference 13).

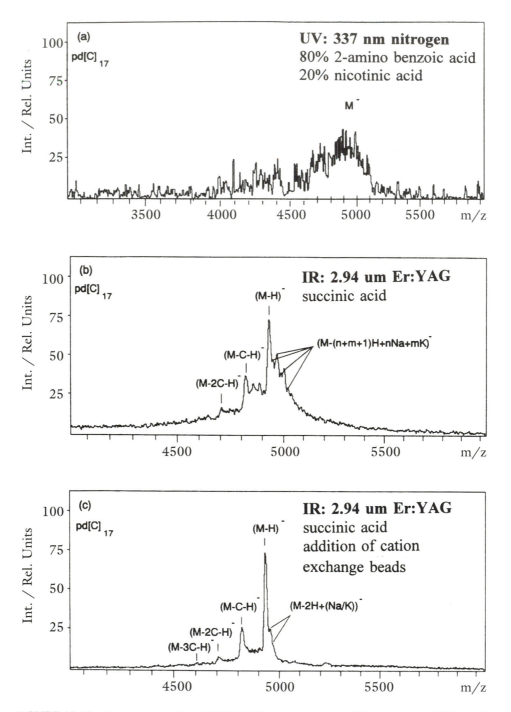

FIGURE 11.12 Comparison of the MALDI–TOF mass spectra of d(C)$_{17}$ using (a) a UV laser, (b) an IR laser, and (c) and IR laser following the addition of cation-exchange beads. (Reprinted with permission from reference 13).

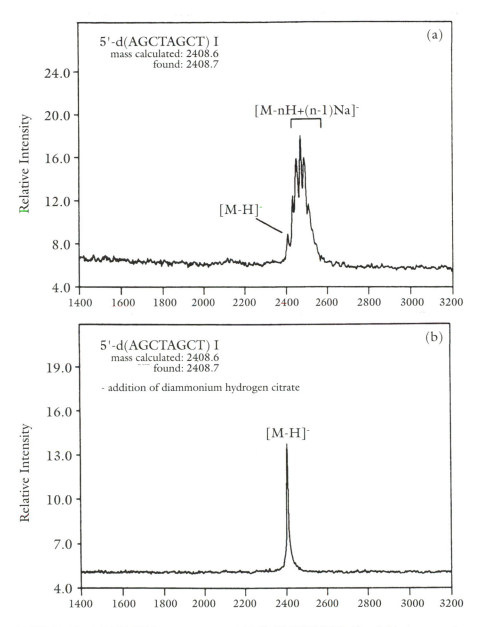

FIGURE 11.13   MALDI–TOF mass spectra of (a) 5′-d(AGCTAGCT)-3′ and (b) the same 8-mer after the addition of diammonium hydrogen citrate. (Reprinted with permission from reference 14).

from ice, which is shown in Figure 11.15. In contrast to the UV laser desorption results for doubly stranded DNA shown in Figure 11.9, the duplex is desorbed intact.[17] In their method, the aqueous matrix is frozen on a copper substrate, which appears to play some role in the absorption of photons from the argon ion laser. Figure 11.16 compares the MALDI mass spectra of a mixture of mixed-base oligomers:

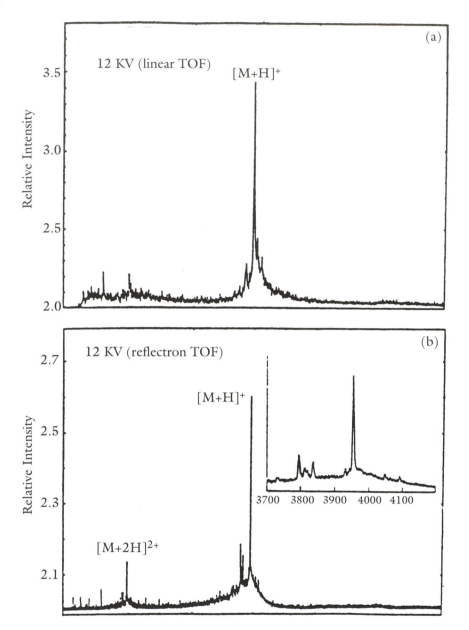

FIGURE 11.14 Comparison of the MALDI–TOF mass spectra of an oligonucleotide obtained on (a) a linear TOF and (b) a reflectron TOF. (Reprinted with permission from reference 15).

8-mer	GACTGACT
10-mer	GACTGACTGT
14-mer	GACTGACTGTGACA
20-mer	GACTGACTGTGACTGACTGT
26-mer	GACTGACTGTGACTGACTGTGATCGT
32-mer	GACTGACTGTGACTGACTGTGACTGTGACTGT
60-mer	GATCGATCGTGATCGATCGTGATCGATCGTGA
	TCGATCGTGATCGATCGTGATCGATCGA

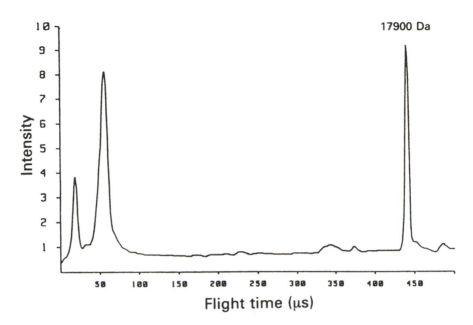

FIGURE 11.15 Single-shot MALDI–TOF mass spectrum of the double-stranded DNA oligonucleotide TrpA Translation Terminator from a frozen aqueous matrix. (Reprinted with permission from reference 17).

obtained using a copper substrate and a permanganate-corroded copper substrate.[18] Both mass spectra show outstanding mass resolution and signal/noise ratios and indicate the technique has considerable promise as an approach to sequencing ladder mixtures.

Sanger Ladder-Sequencing Strategies

As we have noted, the most obvious role for TOF mass spectrometry in sequencing DNA is the sequencing of parallel dideoxy-terminated ladders produced by Sanger synthesis against a complementary strand.[6] Figure 11.17a shows a mock sequencing experiment described by Smith et al.[19] in which mass spectra are recorded for the A, C, G, and T ladders for synthetic nucleotides corresponding to the first 24 DNA fragments that would be generated using a standard M13mp19 template and the primer d(GTAAAACGACGGCCAGT). In Figure 11.17b, these results are overlaid for presentation similar to that obtained by electrophoretic separations. While an impressive early demonstration of this ladder-sequencing approach, considerable advances must be made to reach the 600-mer (or better) range common for electrophoretic methods.

Ladders Generated by Hydrolysis or Exonucleases

Figure 11.18 shows the mass spectrum of the 5′-exonuclease degradation of a DNA 12-mer 5′-d(GCTTXCTCGAGT)–3′ digested with calf spleen phosphodiesterase (CSP).[14] Five peaks are observed, which (if the sequence was unknown)

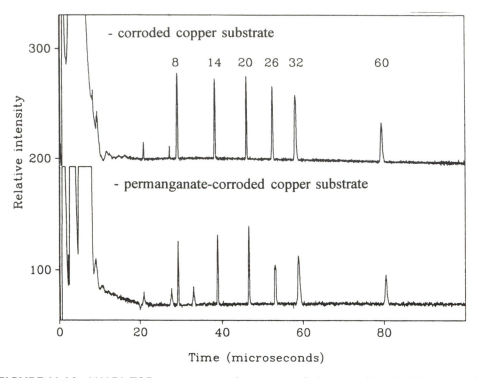

FIGURE 11.16 MALDI–TOF mass spectra of a mixture of oligomers (described in the text) desorbed from a frozen aqueous matrix using an argon-ion laser. (Reprinted with permission from reference 18).

would be used to identify the first four 5′-nucleotide bases from their mass differences. Because the mass differences corresponding to different nucleotides can differ by as little as 9 mass units, this approach places considerably higher demands on mass measurement accuracy than methods using separate dideoxy-terminated ladders. As also shown in Figure 11.18, mass accuracy in this mass range was generally below 1 mass unit. Figure 11.19 shows the mass spectrum of the 3′-exonuclease degradation of the same 12-mer digested with snake venom phosphodiesterase (SVP). The nine peaks observed could be used to identify the first eight 3′-nucleotide bases, and were also measured with less than 1 mass unit accuracy.

An alternative approach to DNA sequencing synthesizes RNA ladders against a DNA template using an RNA polymerase, the nucleotides A, C, U, and G; and 3′-deoxynucleotides as chain-terminators.[20] RNA is attractive for mass-spectral strategies analagous to those used for Sanger DNA ladders, since RNA appears to be more stable to desorption by MALDI. At the same time, mass analysis of RNA by exonuclease sequencing places even more stringent demands on mass accuracy, because it now becomes necessary to distinguish between U and C, which differ by a single mass unit. Nonetheless, Hillenkamp et al.[21] reported MALDI mass spectra for a timed 3′-exonuclease digest of a 54 nucleotide RNA in vitro transcript produced by T7 RNA polymerase that are shown in Figure 11.20. RNA transcripts synthesized by

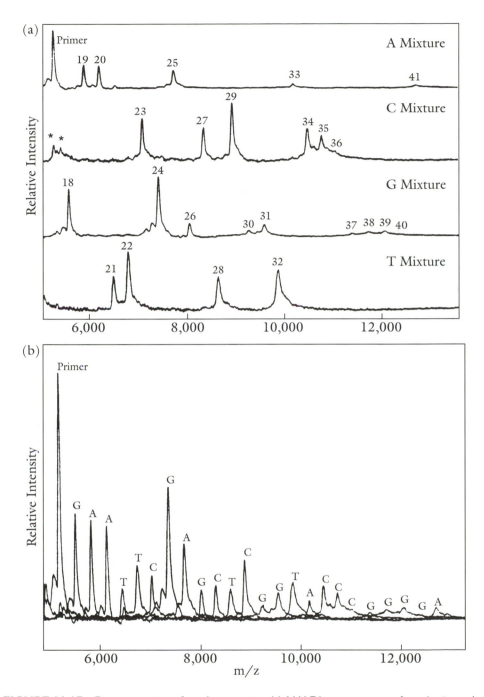

FIGURE 11.17 Demonstration of *mock sequencing*: (a) MALDI mass spectra of synthetic nucleotides corresponding to the first 24 DNA fragments generated in typical sequencing reactions using a standard M13mp19 template and the primer d(GTAAAACGACGGCCAGT) and (b) overlay of the four spectra shown in (a). 0.5 pmol of each component was used and the matrix was 3-HPA. (Reprinted with permission from reference 19).

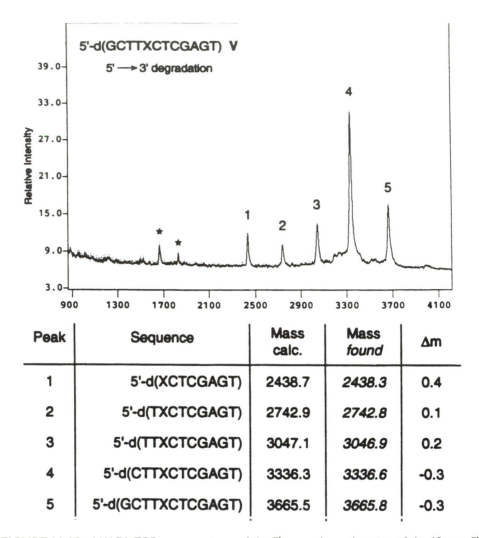

FIGURE 11.18 MALDI–TOF mass spectrum of the 5′-exonuclease digestion of the 12-mer 5′-d(GCTTXCTCGAGT)-3′, and the calculated and measured masses. (Reprinted with permission from reference 14).

Peak	Sequence	Mass calc.	Mass found	Δm
1	5′-d(XCTCGAGT)	2438.7	2438.3	0.4
2	5′-d(TXCTCGAGT)	2742.9	2742.8	0.1
3	5′-d(TTXCTCGAGT)	3047.1	3046.9	0.2
4	5′-d(CTTXCTCGAGT)	3336.3	3336.6	-0.3
5	5′-d(GCTTXCTCGAGT)	3665.5	3665.8	-0.3

bacteriophage RNA polymerases generally carry a 5′-triphosphate group, which can be degraded to a mixture of mono-, di-, and triphosphates that will be unresolved in the mass spectrum and effect mass accuracy. Thus, the intact RNA transcript was treated with calf intestine phosphatase (CIP) to remove 5′-phosphate groups. Additional, nonspecific nucleotides on the 3′-end of the transcript are also common, and contribute to the high mass tailing above the 54-mer peak shown in Figure 11.20a. This 3′-heterogeneity was, of course, removed during the course of the 3′-exonuclease reaction, so that it was possible for them to obtain peaks corresponding to oligomers from the 4-mer to the 54-mer (Figure 11.20c).

Finally, Figure 11.21 shows the IR–MALDI time-of-flight mass spectrum of oligouradylic acids generated by limited hydrolysis of poly-U, using an Er:YAG laser and succininc acid as the matrix.[22]

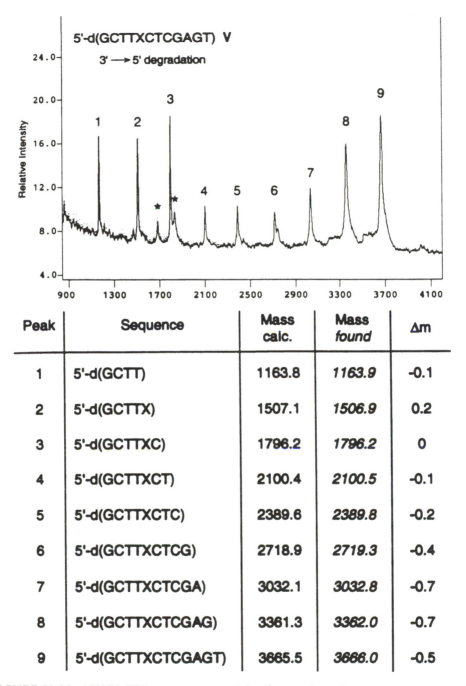

Peak	Sequence	Mass calc.	Mass found	Δm
1	5'-d(GCTT)	1163.8	1163.9	-0.1
2	5'-d(GCTTX)	1507.1	1506.9	0.2
3	5'-d(GCTTXC)	1796.2	1796.2	0
4	5'-d(GCTTXCT)	2100.4	2100.5	-0.1
5	5'-d(GCTTXCTC)	2389.6	2389.8	-0.2
6	5'-d(GCTTXCTCG)	2718.9	2719.3	-0.4
7	5'-d(GCTTXCTCGA)	3032.1	3032.8	-0.7
8	5'-d(GCTTXCTCGAG)	3361.3	3362.0	-0.7
9	5'-d(GCTTXCTCGAGT)	3665.5	3666.0	-0.5

FIGURE 11.19   MALDI–TOF mass spectrum of the 3'-exonuclease digestion of the 12-mer 5'-d(GCTTXCTCGAGT)-3', and the calculated and measured masses. (Reprinted with permission from reference 14).

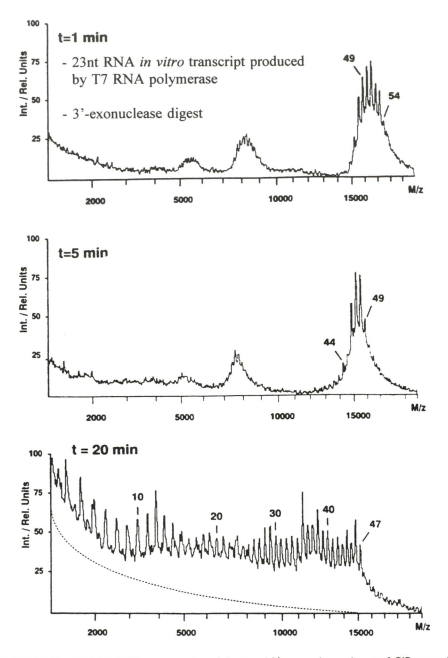

FIGURE 11.20 MALDI–TOF mass spectra of the *timed* 3′-exonuclease digest of CIP-treated 54 nucleotide RNA in vitro transcript produced by T7 RNA polymerase after (a) 1 min, (b) 5 min, and (c) 20 min. (Reprinted with permission from reference 21).

Fragmentation Methods

Martin et al.[23] have recently shown that sequencing information can be obtained from fragmentation of oligonucleotides occuring in the ion source using delayed-extraction MALDI. Figure 11.22 compares the delayed-extraction MALDI mass

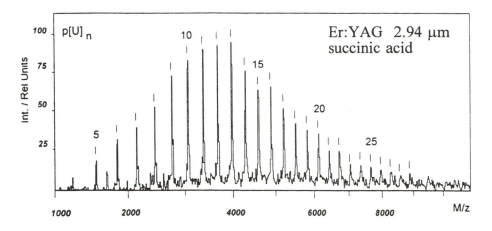

FIGURE 11.21   IR–MALDI mass spectrum of 7 pmol of poly-U following limited hydrolysis. (Reprinted with permission from reference 22).

spectra of the synthetic 11-mer CACACGCCAGT, obtained using two different lasers and matrices. Desorption using a pulsed nitrogen laser (337 nm) and 3-hydroxypicolinic acid as the matrix produced only the molecular (M-H)⁻ and double-charged molecular (M-2H)⁻² ions, at *m/z* 3285 and 1642, respectively. In contrast, desorption using the fourth harmonic from a Nd:YAG laser (266 nm) well above the desorption threshold and picolinic acid as the matrix produced considerable fragmentation. The notation describing these fragments is that suggested by McLuckey et al.[24] and is shown diagrammatically in Figure 11.23. Figure 11.22 also lists these assignments, as well as the calculated and observed masses. Mass accuracies span the range of 0.0 to 3.0 mass units.

Hybridization Strategies

Although the electrophoretic separation of dideoxy-terminated Sanger ladders is the most common method for oligonucleotide sequencing, an important alternative strategy is *sequencing-by-hybridization* (SBH).[25,26] In the SBH approach, an array of all possible (65,536) single-stranded 8-mer probes is immobilized on a chip (Figure 11.24), and exposed to a target solution of fluorescence or radiolabeled, single-stranded DNA fragments having a length (for example) of 3000 bases. The DNA fragments will hybridize to all spots in the array having 8-mer sequences complementary to 8-mer sequences in the fragment. After washing to remove any unhybridized DNA, the spots containing labeled DNA fragments are determined by fluoresence detection or autoradiography. From the set of overlapping 8-mer sequences one then assembles the entire nucleotide sequence. Because the sequences are determined in parallel, SBH is a highly multiplexed method that provides sequencing at high speed. In addition, because all possible overlapping sequences are read, the sequence information is highly redundant, which reduces the error rate. Nonetheless, there are a number of possibilties for errors that include the discrimination

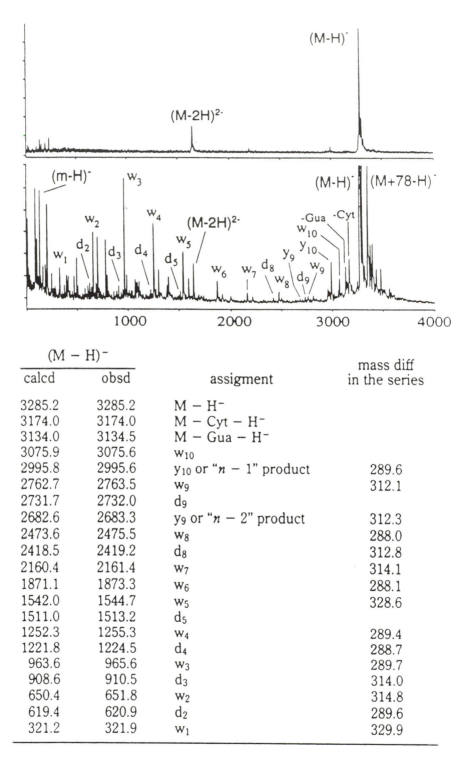

(M − H)⁻			
calcd	obsd	assigment	mass diff in the series
3285.2	3285.2	M − H⁻	
3174.0	3174.0	M − Cyt − H⁻	
3134.0	3134.5	M − Gua − H⁻	
3075.9	3075.6	w_{10}	
2995.8	2995.6	y_{10} or "$n − 1$" product	289.6
2762.7	2763.5	w_9	312.1
2731.7	2732.0	d_9	
2682.6	2683.3	y_9 or "$n − 2$" product	312.3
2473.6	2475.5	w_8	288.0
2418.5	2419.2	d_8	312.8
2160.4	2161.4	w_7	314.1
1871.1	1873.3	w_6	288.1
1542.0	1544.7	w_5	328.6
1511.0	1513.2	d_5	
1252.3	1255.3	w_4	289.4
1221.8	1224.5	d_4	288.7
963.6	965.6	w_3	289.7
908.6	910.5	d_3	314.0
650.4	651.8	w_2	314.8
619.4	620.9	d_2	289.6
321.2	321.9	w_1	329.9

FIGURE 11.22 Delayed-extraction MALDI–TOF mass spectra of the 11-oligonucleotide d(CACACGCCAGT) using (a) a pulsed nitrogen laser and 3-HPA as the matrix and (b) a frequency-quadrupled Nd:YAG laser and picolinic acid as the matrix. (c) Assignment of the fragment ion peaks. (Reprinted with permission from reference 23).

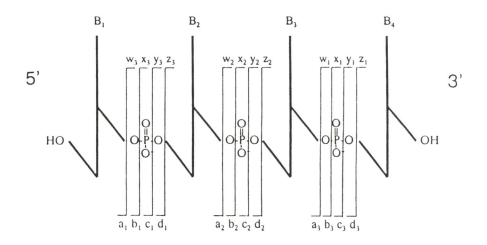

FIGURE 11.23 Notation for the formation of sequence-specific fragment ions for oligonucleotides. (Reprinted with permission from reference 24).

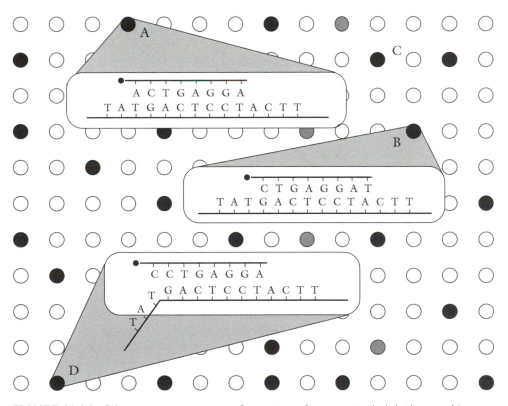

FIGURE 11.24 Schematic representation of a portion of a *sequencing-by-hybridization* chip containing all possible (65,536) 8-mers. Three locations have been expanded to represent the hybridization of an oligomer containing the sequence TATGACTCCTACTT to ACTGAGGA (spot **A**), to CTGAGGAT (spot **B**), and the mismatch to CCTGAGGA (spot **D**).

between correctly matched duplexes and mismatched duplexes, and the existence of repeat sequences.

For example, oligonucleotides containing the internal sequence TATGACTC-CTACTT would hybridize to the location (spot **A** in Figure 11.24) containing the 8-mer ACTGAGGA. This location contains many copies of the same 8-mer probe, and would capture multiple copies of the target DNA fragment. The DNA fragment would also hybridize to the location (spot **B** in Figure 11.24) containing the 8-mer CTGAGGAT, and other locations (spots **C** in Figure 11.24) that enable one to build the entire sequence:

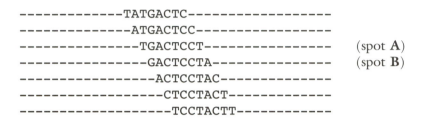

           (spot **A**)
           (spot **B**)

At the same time, this DNA fragment might also hybridize to the location (spot **D** in Figure 11.24) containing the 8-mer CCTGAGGA. Because binding might be expected to be weaker than for a correctly matched duplex, fewer copies of the DNA fragment may be hybridized to this location, resulting in an ambiguous fluorescence or radiography reading. If read as positive, it would incorrectly imply that the DNA fragment contained the sequence GGACTCCT. Conversely, because the binding strength varies with different sequences, a correctly matched duplex might also result in an ambiguous or false negative reading. At the same time, the redundant information provided by overlapping sequences will (in most cases) resolve these problems.

Positional SBH.   Cantor et al.[27] have developed an approach using immobilized duplex (rather than single-stranded) probes, that is intended to provide higher discrimination between false positive (end-mismatched) and false negative (weakly binding, but correctly matched) readings. As shown in Figure 11.25, the duplex probe is composed of an 18-mer hybridized to a complementary 23-mer strand attached to a surface by a noncovalent biotin-streptavidin linkage, resulting in a 5-mer *overhang* in the complementary strand that is used to capture a target DNA fragment with a specific 5-mer 3'-end sequence. Because the position (the 3'-end) of the captured sequence is known, this approach is known as positional sequencing-by-hybridization (positional SBH, or PSBH). After hybridization, the target oligonucleotide fragment is covalently attached to the 18-mer strand using DNA ligase, unless their sequences are displaced or (because they are mismatched) cannot conform to the helical structure imposed by the duplex probe. A *cold wash* then removes any non-ligated fragments, and provides the very high discrimination observed between correctly matched and mismatched sequences. The captured ^{32}P-labeled fragments can then be detected by autoradiography, or released in a *hot wash*.

Mass Spectrometry.   While PSBH provides high discrimination between correctly matched and mismatched oligomers, it detects only end-sequences. However, when used in combination with endonuclease digestion of the DNA fragment (to produce

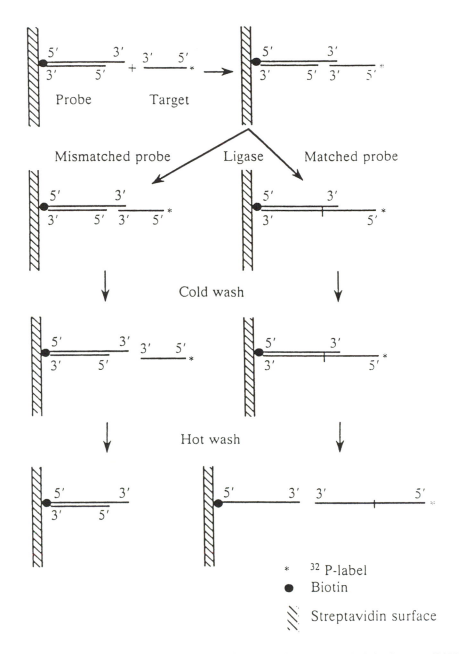

FIGURE 11.25 Schematic representation of *positional sequencing-by-hybridization* (PSBH). (Reprinted with permission from reference 27).

a set of end-sequence targets) and a variety of ladder-generation methods, it can provide a number of effective strategies for mass spectrometric analysis that are well within the current mass range and resolution capabilites of TOF mass spectrometry. In one scheme (Figure 11.26), the DNA fragments would be digested with a suite of restriction enzymes, with each of the resulting fragments captured and ligated to

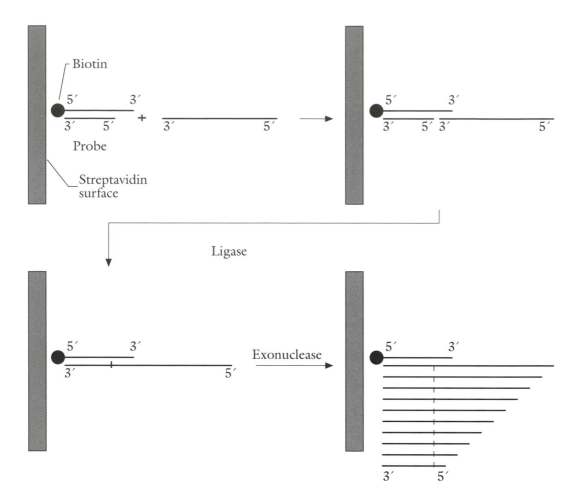

FIGURE 11.26 Scheme for capture, ligation, and exonuclease ladder generation on a duplex probe attached to a streptavidin-coated surface.

one of the 1024 possible 5-mer overhang duplex probes. These would then be digested with 5′-exonucleases to generate ladders that can be analyzed by MADLI mass spectrometry. An important point is that the suite of restriction enzymes can be designed to produce short ladders, so that sequencing could be accomplished by mass difference measurements. For example, if the average maximimum length is 16 nucleotides, then (with the addition of the 18-mer) the mass spectrometer would record oligomers from the 18-mer to 34-mer, well within its capabilities to provide accurate mass measurements. The high sequencing speed would result from the (effectively) simultaneous multiplex recording of oligomer arrays.

In another scheme (Figure 11.27), the same restriction enzyme fragments would again be captured and ligated, but would form templates for extension of the complementary strand using PCR and dideoxy-terminators. Again, if the oligomers are short, then mass difference measurements could be used for a single ladder. Interestingly, the ladder oligomers constructed from either scheme would not have to be removed by a hot wash (as described in Figure 11.25), so that the entire 1024-probe

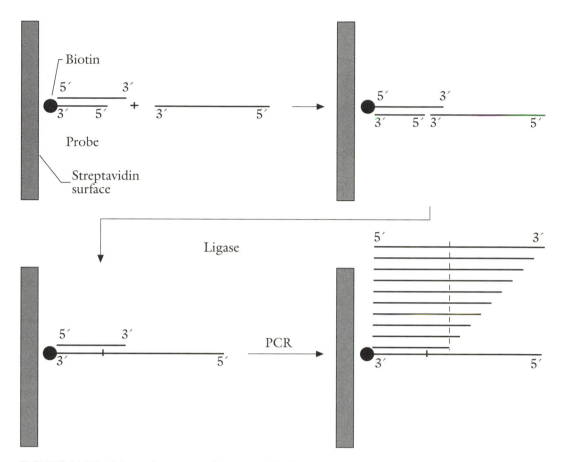

FIGURE 11.27 Scheme for capture, ligation, and ladder generation by polymerase chain reaction (PCR) on a duplex probe attached to a streptavidin-coated surface.

array could be inserted into the mass spectrometer and scanned. For example, Köster et al.[28] have obtained MALDI mass spectra of duplex probes immobilized on streptavidin-coated beads, in which only the set of hybridized synthetic oligomer strands are observed (Figure 11.28).

Thus, mass spectral strategies based upon multiplex recording of short ladder sequences could provide an important alternative to the rather difficult approaches that must necessarily extend the mass range beyond 600-mers (minimally at unit base resolution) to become competitive with existing technologies. In addition, there are opportunities to build considerable redundancy into these strategies, that will insure the correctness of the sequence information. For example, it should be possible to generate overlapping ladders from incomplete digestion by the restriction enzymes. Alternatively, fragments of the initial DNA fragment might be generated by random, controlled hydrolysis, in which case each nucleotide base in the DNA sequence initiates a ladder sequence that will be captured on a specific spot when the solution containing all fragments is exposed to the array. In this case, all of the spots that do not capture a ladder will be used to determine which 5-mer sequences are *not* present in the DNA fragment.

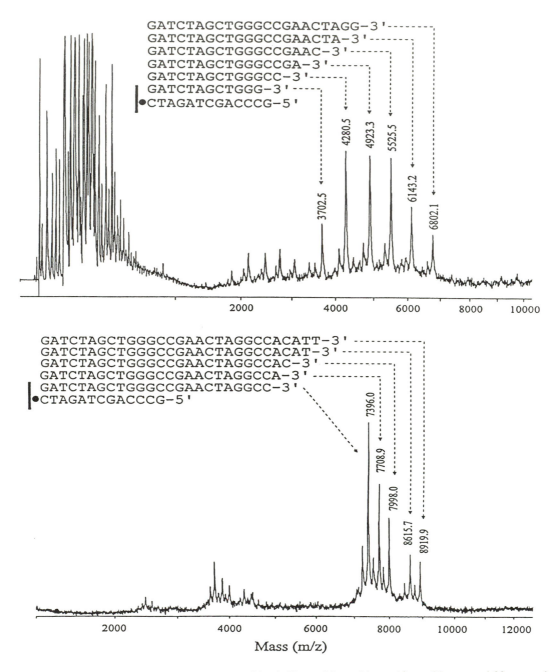

FIGURE 11.28 MALDI–TOF mass spectra of (top) 12mer, 14mer, 16mer, 18mer, 20 mer, and 22mer, and (bottom) 24mer, 25mer, 26mer, 28mer, and 29mer annealed to a 13mer immobilized at the 3'-end to streptavidin-coated magnetic beads. (Reprinted with permission from reference 28).

Finally, as an alternative to the generation of short ladder sequences, captured restriction enzyme fragments could be sequenced from their fragmentation patterns in the mass spectra. This could be accomplished by utilizing delayed extraction to observe in-source decay (as shown in Figure 11.22). Alternatively, fragmentation resulting from post-source decay could be observed in reflectron instruments. In this case, the curved-field reflectron would provide the opportunity to record the product-ion mass spectrum without acquiring multiple spectral fragments at different reflectron voltages, an important opportunity if the instrument is to interrogate 1024 spots for each DNA fragment. In addition, PSD approaches would have a distinct advantage over in-source (and ladder) methods, since it would be possible to mass select from two or more oligomers in the event of a repeat of the captured 5-mer sequence.

References

1. McNeal, C.J.; Ogilvie, K.K.; Theriault, N.Y.; Nemer, M.J., *J. Am. Chem. Soc.* **104** (1982) 972–975.
2. McNeal, C.J.; Ogilvie, K.K.; Theriault, N.Y.; Nemer, M.J., *J. Am. Chem. Soc.* **104** (1982) 976–980.
3. McNeal, C.J.; Ogilvie, K.K.; Theriault, N.Y.; Nemer, M.J., *J. Am. Chem. Soc.* **104** (1982) 981–984.
4. Viari, A.; Ballini, J.-P.; Vigny, P.; Shire, D.; Dousset, P., *Biol. Environ. Mass Spectrom.* **14** (1987) 83.
5. Spengler, B.; Pan, Y.; Cotter, R.J.; Kan, L.-S., *Rapid Commun. Mass Spectrom.* **4** (1990) 99.
6. Sanger, F.; Nicklen, S.; Coulson, A.R., *Proc. Natl. Acad. Sci. USA* **74** (1977) 5463.
7. Schneider, K.; Chait, B.T., *Org. Mass Spectrom.* **28** (1993) 1353.
8. Parr, G.R.; Fitzgerald, M.C.; Smith, L.M., *Rapid Commun. Mass Spectrom.* **6** (1992) 369.
9. Chevrier, M.R., Ph.D. Thesis Dissertation, The Johns Hopkins University, Baltimore, MD (1993).
10. Tang, K.; Allman, S.L.; Chen, *Rapid Commun. Mass Spectrom.* **6** (1992) 365.
11. Tang, K.; Taranenko, N.I.; Allman, S.L.; Chang, L.Y.; Chen, C.H., *Rapid Commun. Mass Spectrom.* **8** (1994) 727.
12. Tang, K.; Allman, S.L.; Chen, C.H.; Chang, L.Y.; Schell, M., *Rapid Commun. Mass Spectrom.* **8** (1994) 183.
13. Nordhoff, E.; Ingendoh, A.; Cramer, R.; Overberg, A.; Stahl, B.; Karas, M.; Hillenkamp, F.; Crain, P.F., *Rapid Commun. Mass Spectrom.* **6** (1992) 771.
14. Pieles, U.; Zürcher, W.; Schär, M.; Moser, H.E., *Nucl. Acids Res.* **21** (1993) 3191.
15. Wu, K.J.; Shaler, T.A.; Becker, C.H., *Anal. Chem.* **66** (1994) 1637.
16. Nelson, R.W.; Rainbow, M.J.; Lohr, D.E.; Williams, P., *Science* **246** (1989) 1585.
17. Nelson, R.W.; Thomas, R.M.; Williams, P., *Rapid Commun. Mass Spectrom.* **4** (1990) 348.
18. Schieltz, D.M.; Chou, C.-W.; Lou, C.-W.; Thomas, R.M.; Williams, P., *Rapid Commun. Mass Spectrom.* **6** (1992) 631.
19. Fitzgerald, M.C.; Zhu, L.; Smith, L.M., *Rapid Commun. Mass Spectrom.* **7** (1993) 895.
20. Axelrod, V.D.; Kramer, F.R., *Biochemistry* **24** (1985) 5716–5723.
21. Kirpekar, F.; Norhoff, E.; Kristiansen, K.; Roepstorff, P.; Lezius, A.; Hahner, S.; Karas, M.; Hillenkamp, F., *Nucl. Acids Res.* **22** (1994) 3866.
22. Nordhoff, E.; Cramer, R.; Karas, M.; Hillenkamp, F.; Kirpekar, F.; Kristiansen, K.; Roepstorff, P., *Nucl. Acids Res.* **21** (1993) 3347.
23. Juhasz, P.; Roskey, M.T.; Smirnov, I.P.; Haff, L.A.; Vestal, M.L.; Martin, S.A., *Anal. Chem.* **68** (1996) 941–946.
24. McLuckey, S.A.; Van Berkel, G.J.; Glish, G.L., *J. Am. Soc. Mass Spectrom.* **3** (1992) 60–70.

25. Bains, W.; Smith, G.C., *J. Theor. Biol.* **135** (1988) 303–307.
26. Drmanac, R.; Drmanac, S.; Strezoska, Z.; Paunesku, T.; Labat, I.; Zeremski, M.; Snoddy, J.; Funkhouser, W.F.; Koop, B.; Hood, L.; Crkvenjakov, R., *Science* **260** (1993) 1649–1652.
27. Broude, N.E.; Sano, T.; Smith, C.L.; Cantor, C.R., *Proc. Natl. Acad. Sci. USA* (1994) 3072–3076.
28. Tang, K.; Fu, D.; Kötter, S.; Cotter, R.J.; Cantor, C.R.; Köster, H., *Nucl. Acids Res.* **23** (1995) 3126–3131.

Summary and Future Prospects

In 1990, Price and Milnes published a review, which they entitled "The Renaissance of Time-of-Flight Mass Spectrometry".[1] This was not (of course) the first prediction of a comeback for the "Toy-Of-Fysicists" (sic). However, it is interesting to note that their review, based upon a paper delivered in July 1989 at the 10th Triennial International Mass Spectrometry Symposium in Salford, concluded with only a brief discussion of two new approaches to laser desorption recently developed by Karas and Hillenkamp[2] and Tanaka et al.[3] that had raised the mass range of the TOF mass analyzer to well above 200 kDa. Thus, the renaissance which they perceived reflected primarily the achievements in biological structural analysis from plasma desorption, surface analysis by laser microprobes, elemental analysis by REMPI instruments, and the use of TOF analyzers in space probes. However, by 1992, scores of laboratories had reproduced the same sensational results from the technique that came to be known as MALDI; and commercial MALDI–TOF instruments for the biological scientist had become available from several manufacturers, including Finnigan (Sunnyvale, CA and Hemmel-Hempsted, UK), VG Biotech (Manchester, UK), Vestec Corporation (Houston, TX), Bruker (Köln, Germany), Linear Instruments (Reno, NV), and Kratos Analytical (Manchester, UK).[4] Still, with the emergence of electrospray ionization and continued concern by mass spectroscopists about the question of mass resolution, it was not at all clear whether this renaissance could be sustained.

As we have seen in this book, three developments appear to have provided the necessary momentum, and all three are still "works in progress". The first was the development of TOF instruments, which could accomodate electrospray ionization by using orthogonal extraction[5] or combining ion trapping with the TOF analyzer,[6] which led to the possibility of carrying out on-line HPLC analysis. Considerable improvements are required to enable ESI–TOF instruments to be commercially competitive with existing quadrupole (and triple quadrupole) ESI mass spectrometers, but this is sure to be a major focus of development by both University laboratories and instrument manufacturers.

The second development was the advent of post-source decay (PSD) methods for peptide amino acid sequencing.[7] While effective strategies had been developed for the structural analysis of peptides and proteins using enzymatic digestion and molecular weight measurements, PSD was clearly a response to the tandem capabilities available using any other mass analyzer. While this means that peptides can now be selectively sequenced from complex mixtures using simple and inexpensive, single-reflectron instruments, the tradeoffs are poor mass selection, reduced mass resolution in the product ion mode, and the need to acquire and calibrate spectra at multiple reflectron voltages. The latter problem has been effectively addressed by the development of a curved-field reflectron,[8] which records the entire product-ion mass spectrum at a single reflectron voltage. However, the inability to mass-select a single monoisotopic species for fragmentation reduces the competitiveness of the reflectron TOF analyzer as a tandem instrument. Thus, one can expect to see the continued development of a variety of hybrid instruments, in which a sector, quadrupole, or ion-trap mass filter provides monoisotopic mass selection.

The third development was the reintroduction of delayed extraction, first described in 1955 by Wiley and McLaren[9] and used in our laboratory for a number of years for focusing ions formed over a broad (up to several microseconds) period.[10,11] A major difference is that delayed extraction is now being used with short-pulsed lasers to achieve impressive high mass resolutions.[12] Additionally, the ability to improve structural information from in-source decay has proven attractive, since it circumvents the need for acquiring and calibrating spectra at many reflectron voltages. However, this approach gives up the important capability for mass selection from mixtures. Ironically (in our view), the technique of post-source decay (PSD) using nonlinear reflectrons should be far more valuable to the protein chemist, but it is the ability to provide very high mass resolutions that has gained the attention (not surprisingly) of mass spectroscopists. However, the technique supports the view that TOF instrument development must indeed strive for high performance, and this can certainly be expected.

Perhaps the clearest indication that the renaissance may at last be upon us is the degree to which TOF instrument development is now being carried out by commercial manufacturers, responding to what appears to be an ever-growing and competitive market. By 1994, PSD instruments using reflectron voltage scanning were available from Finnigan, VG Biotech, Bruker, and PerSeptive Biosystems (Boston, MA), which had recently acquired Vestec. In 1995, Kratos Analytical, which had been acquired by Shimadzu (Kyoto, Japan), introduced a PSD instrument with a curved-field reflectron. VG Biotech also developed and introduced a hybrid sector–TOF instrument using orthogonal extraction of product ions from the collision chamber. In addition, Hewlett-Packard (Palo Alto, CA) entered the TOF arena with its acquisition of Linear Instruments. In 1995, many of these manufacturers introduced delayed extraction instruments, with PerSeptive offering a "high performance" instrument with a 4-m flight tube. At the same time, considerable effort has been made to develop more compact desktop instruments for use by the "non-mass spectroscopist", such as the Kratos Analytical MALDI/I whose price (in 1996) is one third that of the BIO-ION plasma desorption mass spectrometer in 1984! Data systems are now available that include software for comparing mass spectral results (including enzymatic digests) with protein databases avaliable through the Internet, while Ilys Software (Pittsburgh, PA) offers a stand-alone, time-of-flight package (TOFware) for offline processing or for those laboratories that design and construct their own instruments. To date, several hundred MALDI time-of-flight instruments have been sold, with annual sales increasing at a rapid rate. Clearly, the development and popularity of time-of-flight instruments is on an upward swing far steeper than that envisioned in 1990 (and most likely at this time as well).

High Performance Instrumentation for Peptides and Proteins. The analysis of peptides and proteins by MALDI–TOF mass spectrometry is generally routine at this stage. Proteins are extraordinarily well-behaved molecules for mass spectral analysis. With the exception of membrane proteins, their folded (tertiary) structures present a minimal (generally weak and hydrophobic) contact with surfaces or matrices, resulting in facile desorption as intact species with little fragmentation. Detailed structural analysis (including post-translational modifications and amino acid sequences) can generally be obtained from molecular weight measurements of peptides resulting from chemical, endoproteinase, and exopeptidase digests. For the most part, success in mass analyzing peptides depends upon one's ability to obtain peptides with reasonable purity, including the reduction of sodium ions by ion exchange or by the addition of ammonium salts. In this context, one can expect to see improved integration of mass-spectral analysis with biological separations methods, improvments in amino acid sequencing by fragmentation through the development of high performance tandem configurations, and further explorations of non-covalent protein-protein and protein-substrate interactions. Time-of-flight instruments for online HPLC analysis will most likely benefit from further development of ESI–TOF configurations using orthogonal extraction or trapping, but there is considerable need for developing offline mass analysis of peptides and proteins separated by SDS-PAGE and isoelectric focusing, which are more commonly used by protein chemists. The ability to desorb proteins electroblotted onto membranes has already been described,[13,14] so that the real need is for the development of commercial instrumentation that can accomodate intact (2-dimensional) membranes for scanning by the UV laser. Additionally, there is a need to develop more convenient offline MALDI analysis of HPLC, by direct deposition of HPLC fractions onto disposable TOF sample holders.

High-performance, tandem mass spectrometry includes the ability to select monoisotopic molecular-ion species and to efficiently induce fragmentation by collision-induced dissociation. Unit mass selection is decidedly problematic for single-reflectron (RTOF) instruments in which ion gating is carried out in the first linear region prior to mass focusing by the reflectron. Although we have shown that mass selection can be improved by locating the ion gate at the first mass focal point of a tandem (RTOF–RTOF) instrument,[15] unit mass selection is difficult to achieve, since the separation in space between ions differing by one mass unit (for an instrument of any reasonable size) is generally less than the physical dimensions of the gate. In addition, both post-source and in-source fragmentation methods utilize higher laser powers to induce fragmentation, resulting in loss of mass resolution; and the spectra are characterized by a large number of internal ions that make the task of de novo sequencing difficult. Thus, the development of hybrid instruments, using sectors, quadrupoles, or ion traps, can serve to decouple the ionization and fragmentation steps, provide unit mass resolution, and enable better control of fragmentation by the ability to vary collision energies. Sector–TOFs offer the most obvious opportunity for utilizing high-energy collisions, although it has also been shown that the introduction of pulsed, heavy collision gas in quadrupole ion traps can achieve similar results.[16]

Applications to Other Biopolymers. Following the easy success of MALDI–TOF for peptides and proteins, there is now considerable interest in optimizing this technique for other biological polymers, including carbohydrates, glycolipids, and oligonucleotides. For a number of years, infrared laser desorption (IRLD) was used successfully for the analysis of "neutral" polymers, including carbohydrates and industrial polymers such as polyethylene glycol, where ionization occurs by alkali-ion attachment.[17,18] However, analysis by IRLD–TOF mass spectrometry has generally been very limited in mass range, so that there are continued efforts to explore new wavelengths, matrices, and matrix additives that promote cationization, using MALDI–TOF. Electrospray ionization (using quadrupole mass spectrometers) is being used quite sucessfully for the analysis of carbohydrates,[19] so that improvements in ESI–TOF instruments are likely to yield considerable benefits for the growing field of glycobiology. Oligonucleotide analysis by mass spectrometry has always been problematic. MALDI–TOF mass spectrometry has been more successful than previous techniques, but there is considerable need for improvement, particularly for large-scale genomic sequencing. Thus, one can expect that a significant amount of research effort in the next few years will be directed toward developing and improving TOF methods for analyzing other biological and industrial polymers.

Dedicated, Diagnostic, and Miniaturized Instrumentation. A constant theme in the development of new (and softer) ionization techniques has been the ability to mass analyze compounds of increasing molecular size. The time-of-flight mass spectrometer, whose potential for unlimited mass range has been perceived from the beginning, has certainly benefitted from this historical development. At the same time, taking advantage of recent improvements in fast pulsed circuitry and high-speed digitizers, it should be possible to design very compact, low-voltage TOF instruments that would be utilized for the analysis of small molecules. These might include hand-held instruments for drug and explosives detection at airports, chemical detectors in subways, or probes delivered by unmanned rockets for unattended biological warfare agent detection over a period of months or years. While there has always been considerable interest in the development of a "mass-spectrometer-on-a-chip", our laboratory (in conjunction with the Johns Hopkins Applied Physics Laboratory) is focusing upon a more modest and perhaps more practical goal. Based upon previous work on a method for identifying bacteria from their phospholipid signatures,[20] we are currently developing the *TinyTOF* mass spectrometer that utilizes a 3-in. reflectron flight tube, and can achieve unit mass resolution in the mass range of these biomarkers (about 800 Da) using an accelerating voltage of 200 V. The range of possibilities for such miniaturized and dedicated instruments is virtually unlimited, and would also include diagnostic instruments that would be located in the offices or clinical laboratories of physicians. During the next few years, it will be an easy task to demonstrate high performance from low-voltage, compact instruments. The more difficult tasks will involve the development of automated sample preparation, methods for sample introduction into the vacuum system, the maintenance of high vacuum (including miniaturization of pumping systems), and ionization–desorption sources that are considerably more compact than existing short pulsed lasers. Commercialization of such instruments on a wide scale will depend upon these demonstrations and developments, but also upon identifying the specific markets in which they might be employed.

Summary. Following the example of the Price and Milnes report,[1] this volume is also a statement that the renaissance of time-of-flight mass spectrometry is indeed upon us. Begun almost two years ago, my writing of this book has had the advantage of observing the rapid commercialization and utilization of time-of-flight (particularly MALDI–TOF) instruments on a scale that might not have been envisioned in 1990. At the same time, the steep growth curve during the past two years has made the reporting of this subject a most elusive target. Thus, it would not be surprising to find (again) that *we've only just begun.*

References

1. Price, D.; Milnes, G.J., *Int. J. Mass Spectrom Ion Processes* **99** (1990) 1–39.
2. Karas, M.; Hillenkamp, F., *Anal. Chem.* **60** (1988) 2299.
3. Tanaka, K.; Waki, H.; Ido, Y.; Akita, S.; Yoshida, Y.; Yoshida, T., *Rapid Commun. Mass Spectrom.* **2** (1988) 151.
4. Cotter, R.J., *Anal. Chem.* **64** (1992) 1027A.
5. Dodonov, A.F.; Chernushevich, I.V.; Laiko, V.V., in *Time-of-Flight Mass Spectrometry,* Cotter, R.J. (Ed.); ACS Symposium Series 549, Washington, DC (1994) pp 108–123.
6. Cien, B.M.; Lubman, D.M., *Anal. Chem.* **66** (1994) 1630–1636.
7. Kaufmann, R.; Spengler, B.; Lutzenkirchen, F., *Rapid Commun. Mass Spectrom.* **7** (1993) 902–910.
8. Cornish, T.J.; Cotter, R.J., *Rapid Commun. Mass Spectrom.* **8** (1994) 781–785.
9. Wiley, W.C.; McLaren, I.H., *Rev. Sci. Instr.* **26** (1955) 1150–1157.
10. Tabet, J.-C.; Cotter, R.J., *Anal. Chem.* **56** (1984) 1662.
11. Olthoff, J.K.; Honovich, J.P.; Cotter, R.J., *Anal. Chem.* **59** (1987) 999–1002.
12. Vestal, M.L.; Juhasz, P.; Martin, S.A., *Rapid Commun. Mass Spectrom.* **9** (1995) 1044–1050.
13. Strupat, K.; Karas, M.; Hillenkamp, F.; Eckerskorn, C.; Lottspeich, F., *Anal. Chem.* **66** (1994) 649–658.
14. Vestling, M.; Fenselau, C., *Anal. Chem.* **66** (1994) 471–477.
15. Cornish, T.J.; Cotter, R.J., *Org. Mass Spectrom.* **28** (1993) 1129–1134.
16. Doroshenko, V.M.; Cotter, R.J., *Anal. Chem.* **68** (1996) 463–472.
17. Spengler, B.; Dolce, J.W.; Cotter, R.J., *Anal. Chem.* **62** (1990) 1731–1737.
18. Cotter, R.J.; Honovich, J.P.; Olthoff, J.K.; Lattimer, R.D., *Macromolecules* **19** (1986) 2996.
19. Duffin, K.L.; Welply, J.K.; Huang, E.; Henion, J.D., *Anal. Chem.* **64** (1992) 1440–1448.
20. Platt, J.A.; Uy, O.M.; Heller, D.N.; Cotter, R.J.; Fenselau, C., *Anal. Chem.* **60** (1988) 1415–1419.

Index